SCI-MATH

Applications in Proportional Problem Solving

Module Two

Madeline P. Goodstein

Addison-Wesley Publishing Company
Menlo Park, California   Reading, Massachusetts
London   Amsterdam   Don Mills, Ontario   Sydney

ADVISORY COMMITTEE

Sigmund Abeles
State Science Consultant
Connecticut State Department of Education

Max Bell
Mathematics Education
University of Chicago

Steven Detweiler
Department of Physics
Yale University

J. Dudley Herron
Department of Chemistry
Purdue University

William M. Masterton
Department of Chemistry
University of Connecticut

Donald Palzere
Department of Applied Mathematics
Central Connecticut State College

This book is published by the ADDISON-WESLEY INNOVATIVE DIVISION.

Design by Suzanne Bonfield.
Illustrations by Jane McCreary.

This material is based upon research supported by the National Science Foundation under Grant No. SED77-18444. Any opinions, findings, and conclusions or recommendations expressed in this publication are those of the author and do not necessarily reflect the views of the National Science Foundation.

Copyright © 1983 by Addison-Wesley Publishing Company, Inc. All rights reserved. No part of this publication may be reproduced, stored in a retrieval system, or transmitted, in any form or by any means, electronic, mechanical, photocopying, recording, or otherwise, without the prior written permission of the publisher. Printed in the United States of America. Published simultaneously in Canada.

Except for the rights to materials reserved by others, the publisher and the copyright owner hereby grant permission without charge to domestic persons of the United States and Canada for use of this Work and related materials in the English language in the United States and Canada after January 1, 1988. For conditions of use and permission to use the Work or any part thereof for foreign publications or publications in other than the English language, apply to the copyright owner or the publisher. Publication pursuant to any permission shall contain an acknowledgment of this copyright and an acknowledgment and disclaimer statement as shown above.

ISBN-0-201-20074-0
ABCDEFGHIJKL-ML-898765432

Preface to Students

SCI-MATH is designed to help prepare you to be able to do the calculations you need in the introductory sciences. The same mathematical techniques are also useful for problem solving in business, consumer purchasing, industry, crafts, and even in the home.

SCI-MATH focuses on the understanding of the *concept* of proportions and on the *use* of proportions in problem solving. In essence, the mathematics taught is the mathematics needed for science. It is also the basis for understanding the real-world applications of fundamental equations in algebra and in parts of geometry.

The **SCI-MATH** curriculum is divided into two modules:

Module One A pre-algebra module dealing with the arithmetic and logic of proportions.

Module Two An algebra module that examines the symbolic equations and graphing of proportions.

Acknowledgments

This text was developed with the support of the National Science Foundation. Without such support, the time necessary to write these materials would never have been available. Moreover, the subsequent testing and revisions of the text which critically improved its effectiveness would not have been possible. For all those who use any part of this text or its ideas, a debt of gratitude is owed to the National Science Foundation.

The author would like to express thanks and appreciation to the many people who helped to initiate this project and/or generously gave their time to improve the text materials:

to *Mary O'Brien*, Science Coordinator, Board of Education of Milford, Connecticut, who, with high courage and sensitivity to the needs of her charges, made Foran High School available for our first field-test.

to the administrators, teachers, and students of the ten high schools in Connecticut who participated in the field-tests of our materials, and especially to the following:

Lawrence F. Bencivengo, North Haven High School, North Haven
Fradelle Fuhrman, Berlin High School, Berlin
Jane Gwodz, RHAM High School, Hebron
Joseph Kashman, Foran High School, Milford
Lawrence Kershnar, New Milford High School, New Milford
George Lelievre, East Windsor High School, East Windsor
Rose Paternostro, Newington High School, Newington
Laura Pels, Bulkeley High School, Hartford
Albert Pesticci, North Haven High School, North Haven
William Pilotte, Newington High School, Newington
Gerald Robbins, North Haven High School, North Haven
Brother Thomas Sawyer, Notre Dame High School, West Haven
Laurence I. Tripp, New Milford High School, New Milford
Lester E. Turner, Hillhouse High School, New Haven

Thanks and appreciation are also due to *Helen Flis*, Grants Accountant at Central Connecticut State College, who never failed to respond precisely, rapidly, and courteously to every request made of her office, and to the many administrators and staff members at Central Connecticut State College who aided in carrying out this grant. Thanks are extended to *Linda Kahan* and *Mary Ann Ryan* of the National Science Foundation whose consideration and responsiveness were so helpful. *Isabel Fairchild* made the drawings for our second field-test edition; she showed the delicate humor of mathematics in her delightful line drawings. I would also like to express my appreciation to *Dr. William Boelke* of the Department of Applied Mathematics at Central Connecticut State College who was the Assistant Project Director during the second phase of this project in charge of the statistical analyses of data; he freely gave his time, expertise, and helpful advice in treating the data during both phases of the project. Thanks also go to *Prof. Donald Palzere* for his help, and to *Prof. J. Dudley Herron* for his invaluable advice.

Finally, I want to thank my husband, *Julian Goodstein*, without whose continued encouragement and patience these materials would never have been completed.

ACKNOWLEDGEMENTS

Contents

Chapter Three
HOW TO WRITE AN EQUATION FOR A DIRECT PROPORTION

Chapter Four
INVERSE PROPORTIONS—the constant product or seesaw relationship

1

Relationships

1.1 INTRODUCTION

Algebra is a language of symbols whose rules enable all kinds of calculations to be carried out using the symbols. When the symbols describe real things like the price of clothing, the speed of a car, or the distances between planets, algebra can help to simplify calculations about these things and, more important, it can help us to understand the relationships between these things so that we can make predictions about them. This is where the real power is, to make predictions. It is our modern form of magic, to make predictions that turn out to be true.

Although we can predict with fair success such things as profits, birth rates, food consumption, employment and other human activities, it is in the field of science that we are truly like magicians. The time of an eclipse, the direction of a projectile, the amount of chemical produced in a reaction, the growth of an animal, and much much more can all be precisely foretold. Mathematics makes it possible for scientists to do this, and it all begins with the use of algebra in the real world.

Our purpose in this book is to examine the numerical connections between things in nature, to state these in symbolic equations, and to see how quantities can be predicted. Before you have completed this book, you will yourself have written some equations to describe numerical relationships between real things.

1.2 DEFINITIONS

Before we proceed to the study of how algebraic equations can be put to practical use, let us review the idea of relationships between quantities of variables—quantitative relationships. If you have not used Module One of this series, some of the ideas here may be new to you. Otherwise, this will serve to help you recall the ideas.

We are going to be talking about quantities of variables and constants so let us define these terms first.

Quantity: A quantity of something tells how much there is of it, and is always expressed by a number and the unit of measurement. Examples of quantities are ten pens, $500, one million records, 5 fingers, or 3/4 pound of potatoes. A quantity usually has to be measured. Note that counting is a form of measurement.

Variable: Variables are things that can change in quantity depending upon the situation, such as the quantity of ink left in your ball point pen, the amount of money spent on lunch in your school today, or the total number of copies of a hit record sold so far.

Constant: Things that are not variable are constant; they are fixed in quantity and don't change. Examples of constants are the number of fingers on a normal hand, the number of inches in a foot, and the speed of light in a vacuum.

In abstract algebra, numbers without units may be assigned as the values for constants and variables. In the real world, however, measurements are usually needed to describe how much there is of a variable or constant, and so each is expressed as a number with its measurement unit.

1. Identify the quantities from among the following:
   a. 76 cents
   b. $\frac{71}{5}$
   c. 6 things
   d. houses
   e. 7 cm
   f. 2 quarts

In Problems 2 to 8, state which ones are variables:

2. Outdoor temperature.
3. Number of cents in one dollar.
4. Number of pennies in a supermarket register.
5. Weight of a measuring cup of solid margarine.
6. Length of a baby.
7. Gasoline in the tank of a moving car.
8. Barometric pressure.

1.3 RELATIONSHIPS

Let us start with the idea of two variables that have a *relationship* to each other, that is, *when one changes, so does the other.* For example, the number of gallons of gasoline and the money you pay for it are variables that are connected; they have a relationship to each other.

If one variable gets bigger when the other gets bigger, or if it becomes smaller when the other becomes smaller, it is called a *direct relationship*. The two variables vary directly with each other. The more gallons of gasoline that you buy, the more you must pay.

$A\uparrow \quad B\uparrow \qquad \text{or} \qquad A\downarrow \quad B\downarrow$

There are also certain variables where one gets bigger when the other gets smaller or becomes smaller if the other becomes bigger, such as the number of hours of daylight and the number of hours of nighttime in one full week. These are called *inverse relationships*: another name for this is the *indirect relationship.*

$C\uparrow \quad D\downarrow \qquad \text{or} \qquad C\downarrow \quad D\uparrow$

We shall start by considering a special case of the direct relationship called the *direct proportion.* When two variables are connected by a direct proportion, *if one of them changes by any factor, the other changes by the same factor.* Numbers used to multiply are called *factors*; another word for factor is *multiple*. For example, if one variable changes by a factor of four, that is, if it becomes four times as big as it was before, then the other variable is directly proportional if it also changes by a factor of four to become four times as big as it was before. If you spend four times as much for gasoline, you will buy four times as much.

Likewise, if you spend only 1/5 as much for gasoline as you did before, you will only get 1/5 as much gasoline.

If the factor of change is bigger than one, then both variables change to become bigger by the same multiple.

If the factor of change is less than one, then both of the variables become smaller. They become the same fraction of what they were before.

Suppose we consider the variables of distance traveled and volume of gasoline. Are these two variables proportional to each other? We know that, for example, if a car travels 30 miles for one gallon of gasoline, we will expect it to drive 60 miles for two gallons (factor of 2), or fifteen miles for one-half gallon (factor of ½). Evidently, these two variables are proportional to each other. There are several ways to express this proportion mathematically. One way is to write

$$\frac{30 \text{ miles}}{60 \text{ miles}} = \frac{1 \text{ gallon}}{2 \text{ gallons}}, \text{ or } \frac{60 \text{ miles}}{30 \text{ miles}} = \frac{2 \text{ gallons}}{1 \text{ gallon}}.$$

Both of the above say the same thing. In both cases, the units can be cancelled out to give

$$\frac{30}{60} = \frac{1}{2} \text{ or } \frac{60}{30} = \frac{2}{1}.$$

These are called *ratio proportions* because each side of the equation consists of a ratio.

A ratio is a dimensionless quantity; it has no measurement units in it, or else the measurement units can be canceled out. Thus,

$$\frac{30}{60}, \frac{1}{2}, \frac{60}{30}, \text{ and } \frac{3 \text{ gallons}}{1 \text{ gallon}}$$

are all ratios.

We can also show the proportional relationship between distance and volume of gasoline with a different kind of equation:

$$\frac{60 \text{ miles}}{2 \text{ gallons}} = \frac{30 \text{ miles}}{1 \text{ gallon}}.$$

We can get this equation by rearranging the ratio proportion. The 30 miles is now directly connected to the 1 gallon. We can also write

$$\frac{60 \text{ miles}}{2 \text{ gallons}} = \frac{30 \text{ miles}}{1 \text{ gallon}} = \frac{15 \text{ miles}}{\frac{1}{2} \text{ gallon}} = \frac{90 \text{ miles}}{3 \text{ gallons}}$$

and so on. Writing the proportion in this manner extends the way we can look at the proportional relationship.

The quotient of an ordered pair of variables is called a *rate*. A rate *tells how much of one variable there is for a quantity of a second variable.* It tells us *how much per.* An expression that equates two rates, such as

$$\frac{90 \text{ miles}}{3 \text{ gallons}} = \frac{30 \text{ miles}}{1 \text{ gallon}} \quad \text{or} \quad \frac{60 \text{ miles}}{2 \text{ gallons}} = \frac{600 \text{ miles}}{20 \text{ gallons}}$$

is called a *rate proportion.*

Not all rates form proportions. For example, a pad of paper may cost 40¢/pad but costs 70¢/2 pads. (Notice that we customarily leave out the "one" in an expression like 40¢/pad.) When variables are proportional to each other, the rates that connect them are called *invariant.* All the expressions of an invariant rate are equivalent and can be reduced to equal each other. The rate of 12 cans lemonade/case is invariant:

$$\frac{12 \text{ cans}}{\text{case}} = \frac{18 \text{ cans}}{1\frac{1}{2} \text{ cases}} = \frac{24 \text{ cans}}{2 \text{ cases}} = \frac{120 \text{ cans}}{10 \text{ cases}} \ldots$$

When the rate is invariant, the two variables are proportional to each other.

The two forms of a proportion, the ratio proportion and the rate proportion, give us two keys to identify when a relationship is a direct proportion.

1. *When one variable changes, the other changes by the same factor.* Hence, the ratio of the new to old quantities is the same for each variable. The ratio is the factor of change.

$$\frac{90 \text{ miles}}{30 \text{ miles}} = \frac{3 \text{ gallons}}{1 \text{ gallon}}.$$

2. *The rate connecting the two variables is invariant.*

$$\frac{90 \text{ miles}}{3 \text{ gallons}} = \frac{30 \text{ miles}}{1 \text{ gallon}} = \ldots.$$

Answer the following:

1. Which of the following rates is invariant?
   a. Each dress in this store is $19.75.
   b. 5 lbs./$1, 25¢/lb. if less than 5 lbs. is purchased.
   c. 2 for $1, 59¢ each.
   d. Twelve inches per foot.
   e. John ran the first mile in 6 minutes, the first two miles in 12 minutes, and the first three miles in 20 minutes.

2. State whether the variables in each of the following have a direct relationship, an inverse relationship, or probably do not have any relationship:

   a. Number of slices eaten in a loaf of bread and number of slices remaining.

   b. Amount of margarine by volume measured in tubs, quarts, cups, litres, etc. and weight of margarine.

   c. Lengths of a racetrack and time required to run them at the same speed.

   d. Speed of a vehicle and time required to cover one mile of distance.

   e. The weight of a car and how dark its color is.

   f. Electric bill per month and number of hours of daylight.

3. In the following, identify which are ratios:

   a. 1000 gallons/mile.

   b. 5 policemen/town.

   c. 5 policemen/10 policemen.

   d. 6/3.

4. Convert the following rate equations to ratio equations:

   a. $\frac{55 \text{ miles}}{\text{hr.}} = \frac{110 \text{ miles}}{2 \text{ hrs.}}$.

   b. $\frac{2 \text{ quarts oil}}{\text{car}} = \frac{10 \text{ quarts oil}}{5 \text{ cars}}$.

   c. $\frac{25 \text{ balls hit}}{100 \text{ balls pitched}} = \frac{65 \text{ balls hit}}{260 \text{ balls pitched}}$.

5. State the factor of change for each equation in the preceding Problem 4.

6. Rearrange the following ratio proportion to a rate proportion. Also, state the factor of change.

   $\frac{24 \text{ eggs}}{3 \text{ eggs}} = \frac{2 \text{ dozen}}{\frac{1}{4} \text{ dozen}}$.

7. Which of the following tables describes a direct proportion?

a.

X (hrs.)	Y (feet)
1	5
2	10
3	15
5	25

b.

g	r
1	5
2	10
3	20
4	40
5	80

c.

A	B
11	6
17	12
6	1
20	15

d.

Q	R
10	5
20	10
16	8
8	4

1.4 THE UNITARY RATE

The use of rates is very convenient when working with variables that have a proportional relationship. The following rule is especially useful when working with an invariant rate.

> ***If you know how much per one of any quantity, then you can calculate how much per any multiple of that quantity.***

For example, if you know how much per one pound of cake, you multiply the rate by five to find out how much per 5 lbs. of cake, or by 50 to find out how much per 50 pounds of cake, or by ½ to find out how much per ½ pound of cake.

We have a special name for the rate that tells how much *per one* of a thing. It is called the *unitary rate.* The rule above is called the *Unitary Rate Rule*.

Use the Unitary Rate Rule to solve the following. It is not necessary to reduce a given rate to the unitary rate to apply the Unitary Rate Rule since all the rates for any given proportion are equivalent to the unitary rate and can be substituted for it.

1. Fill in the missing quantities in the following rate proportions. Apply the Unitary Rate Rule:

   a. $\dfrac{\text{3 tennis balls}}{\text{can}} = \dfrac{\square}{\text{5 cans}}$.

   b. $\dfrac{\text{24 cans soft drink}}{\text{case}} = \dfrac{\square}{\text{14 cases}}$.

   c. $\dfrac{\$10{,}750}{\text{25 suits}} = \dfrac{\square}{\text{18 suits}}$.

Solve all of the following problems by applying the Unitary Rate Rule.

2. How far will 300 gallons take a car that gets 20 miles to the gallon?

3. How many gallons are needed to drive a car that gets 20 miles to a gallon a distance of 300 miles?

4. The density of a certain gas is 1.400 g per litre. Find the mass of 0.0500 litres of the gas.

5. A sewing machine is adjusted to sew 35 stitches per 3 inches. How many stitches are needed to sew 136 inches?

6. Sixteen cans of oil are needed for 5 machines. How many cans are needed for 8 machines? Prove that you are using the Unitary Rate Rule.

7. A box of pencils has a mass of 120 g. What is the mass of 126 boxes?

8. What is the cost of 0.5 quarts of milk at 3 quarts for one dollar?

9. What is your score if you got 17 correct out of 75 questions on a hundred-point test?

10. A photograph of Eric and Alice shows Eric 12 cm high and Alice 10 cm high. Eric's height is actually 182 cm. What is Alice's real height?

1.5 CALCULATIONS WITH MEASUREMENT UNITS

The measurement units that are part of quantities should not be left out of the calculations that you do with real-world and scientific quantities. Measurement units can be added, subtracted, multiplied and divided according to the rules in this section.

A. Addition and Subtraction with Units

The rule for addition of numbers with units is:

> ***To add numbers with units, only quantities with the same kind of unit may be added. To add quantities with the same unit, add up all of the numbers and give the total the same unit.***

The rule for subtraction is similar and is:

> ***Only quantities with the same units may be subtracted from each other. Assign the same unit to the difference.***

A unit is the same whether singular or plural, so inch and inches are the same unit, foot and feet are the same unit, and meter and meters are the same unit. As an example, note that one foot plus two feet equals three feet; these quantities may be added together because they have the same unit.

Now, let us consider one more thing about addition and subtraction of quantities. What is the sum of 3 baseball bats, 2 tennis rackets, 5 baseball bats, and another tennis racket? The sum is 8 baseball bats and 3 tennis rackets. Since bats and rackets are different variables, they can't be added together. If, however, bats and rackets were to be combined under a *new* name such as pieces of athletic equipment, then and only then could they all be considered the same variable and added together to give 11 pieces of athletic equipment.

Apply the above rules to solve the following problems.

1. Given a box with 6 baseballs and two tennis balls. Five more baseballs and a dozen tennis balls are tossed into the box. What does the box now contain?

2. Given three jewel boxes that *each* contain two emeralds, a golden apple, and five sapphires.

   a. Which variable has a total of six present?
   b. Which variable totals seven in each box?

3. What is the sum of 12 inches, 2 feet, and 12 feet?

4. How much is 22 inches subtracted from two feet?

5. A box contains 150 green jelly beans, 30 red jelly beans, and 20 black jelly beans. How many jelly beans are in the box?

6. What is the sum of 4 yards, 4 feet, and 4 inches?

7. Invent an example of subtraction of quantities and state the answer.

B. Multiplication and Division of Units

Since multiplication is a form of addition, it follows that if we perform the operation,

3×4 yards,

the answer is 12 yards; this is equivalent to adding 4 yards three times, that is, 4 yards + 4 yards + 4 yards.

Suppose, however, that both numbers are quantities such as:

3 yards $\times$ 4 yards.

The rules for the operations on the units in the above calculation are quite similar to but not identical to those for multiplication and division of numbers. There is no specific name for operations with the units parts of quantities, so let us call them unit-multiplication and unit-division, keeping in mind that these terms are not the same as ordinary multiplication and division with numbers.

Rather than describe the rules in detail with words, it is simpler to show them by examples as follows.

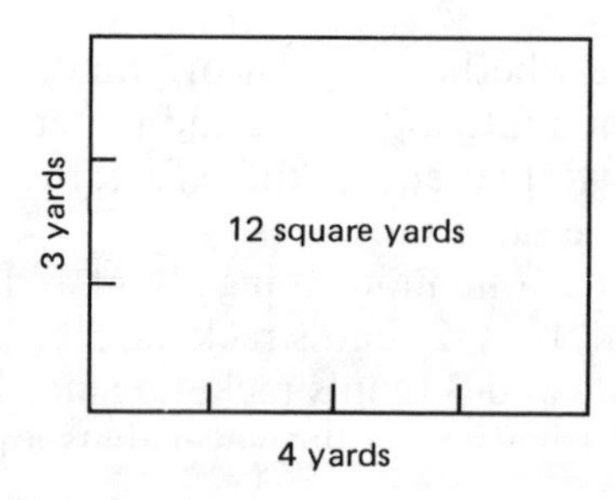

Unit-multiplication

(1) 3 yards × 4 yards = 12 yards × yards or
12 yards · yards or
12 yards^2 or
12 square yards

(2) 3 yards × 4 yards × 2 yards = 24 yards × yards × yards
or 24 yards · yards · yards
or 24 yards^3
or 24 cubic yards

Unit-division

(1) $\dfrac{5 \text{ yards}}{2 \text{ rooms}} = 2.5 \dfrac{\text{yards}}{\text{room}}$

(2) $\dfrac{6 \text{ yards}}{6 \text{ yards}} = 1$

(3) $\dfrac{5 \text{ yards}}{6 \text{ yards}} = \dfrac{5}{6}$

Unit-multiplication and Unit-division

(1) $\dfrac{(12 \text{ yards})\ (2 \text{ quarts})}{3 \text{ yards}} = 8 \text{ quarts}$

(2) $\dfrac{(12 \text{ yards}^2)\ (3 \text{ rooms})}{4 \text{ yards}} = 9 \text{ yards} \cdot \text{rooms}$

You can unit-multiply and unit-divide by following the above examples even though the meanings of some or all of the units may not be known to you. For example, consider the calculation, 6 erasers × 6 pencils = 36 erasers · pencils. A unit such as eraser · pencil probably has no useful physical meaning. However,

$$\frac{6 \text{ erasers}}{6 \text{ pencils}} = \frac{1 \text{ eraser}}{\text{pencil}}$$

has a physical meaning. It says that if there are six erasers per six pencils, then there is one eraser per pencil. Thus, we see that when we are working with real life variables, only those unit-multiplication and unit-divisions that can express real meanings are useful to us.

Solve the following problems; do not be concerned at this point about the meaning of the units.

8. 6 kilowatts × 2 hours
9. 5 cm × 5 cm
10. 5 cm × 5 cm × 5 cm
11. 36 miles ÷ 6 hours
12. 100 people ÷ 2 rooms
13. (60 cm) (cm) ÷ 30 cm
14. 40 cm^2 ÷ 20 cm
15. 60 g × 3 cm
16. (60 g) (3 cm) ÷ 5 sec
17. $\dfrac{39¢ \times 6 \text{ apples}}{2 \text{ apples} \times 2 \text{ children}}$

C. Equations with Units

When any equation has quantities in it, both numerical value and units must be equal on both sides of the equation. The following exercises require unit-multiplication and/or unit-division to make the units equal on both sides of the equations. To concentrate attention on the units, the numbers have been omitted.

18. gallons × $\dfrac{\text{tanks}}{\text{gallons}}$ = ☐.
19. boards × $\dfrac{\text{feet}}{\text{board}}$ = ☐.
20. $\dfrac{\text{centimeters}}{\text{second}}$ × seconds = ☐.
21. miles × $\dfrac{\text{hours}}{☐}$ = hours.
22. dollars × $\dfrac{☐}{\text{dollar}}$ = cans soda.
23. ☐ × $\dfrac{\text{letters}}{\text{inch}}$ = letters.
24. jackets × $\dfrac{☐}{☐}$ = buttons.
25. $\dfrac{☐}{☐}$ × boxes = staples.
26. $\dfrac{\text{kg} \cdot \text{m}^2}{\text{sec}^2}$ × ☐ = kg.
27. Given that $F = \dfrac{kQ^2}{r^2}$ and that Q is measured in coulombs, r is measured in meters, and F is measured in newtons, state the units for k.

1.6 RATES AND RECIPROCAL RATES

There is one more aspect of rates to consider before beginning the study of the algebra of relationships.

Consider the rate, $\dfrac{25 \text{ students}}{\text{classroom}}$. This tells us that there are 25 students per classroom; it follows, then, that there is one classroom per 25 students. These two rates may be written as, respectively,

$$\frac{25 \text{ students}}{\text{classroom}} \quad \text{and} \quad \frac{1 \text{ classroom}}{25 \text{ students}}.$$

These two rates are said to be reciprocals* of each other, that is, 25 students/classroom is the reciprocal of 1 classroom/25 students and, likewise, 1 classroom/25 students is the reciprocal of 25 students/classroom. By inverting a rate, its reciprocal is obtained. Answer the following problem.

1. For each of the following, state the reciprocal rate.

    a. 2000 staples/box
    b. 1 gallon/$1.35
    c. 24 hours/day
    d. 88 keys/piano

When a rate and its reciprocal rate are both converted to unitary rates, some interesting comparisons may sometimes appear. For example, if you earn $200 per week, the reciprocal of this is 0.005 week/$ which means that you need only work $\frac{5}{1000}$ of a week to earn a dollar. Answer the following.

2. State the unitary reciprocal rate of each of the following.

    a. $1 / 3 lbs.
    b. 24 hours / day
    c. 25 miles / gallon of gasoline
    d. 25¢ / can of soda pop

We see that any relationship described by a rate can be expressed in two ways. For example, if a car moves at 50 miles/hour, then it also moves at 1 hour/50 miles or 0.02 hour/mile. If a person earns $400 per week, then it takes 0.0025 weeks to earn $1 or 0.0025 week/dollar. Thus, each such relationship can be expressed by a rate or the reciprocal of that rate.

3. Express each of the following by two unitary rates, one of which is the reciprocal of the other:

    a. 5 airplanes for each hangar.

    b. 64 women per 36 men.

    c. 44 grams carbon dioxide per mole of carbon dioxide.

    d. For every year, there are 365 days.

    e. 24 clips to 32 fasteners.

In the remainder of this module, *include the measurement units in all of your calculations.* Measurement units can be used to help analyze the nature of a variable, to check that a problem is solved correctly, and even to help set up the procedure for solving a problem. When used in any of these ways, the process is called units analysis. Other names for the process of units analysis are: dimensional analysis, factor-label method, and quantity calculus. See Module One of this series for lessons on how to carry out simple units analysis.

*Another term for reciprocal is multiplicative inverse.

DEFINITIONS

Constant: A number or quantity that does not change.

Variable: A variable is something that can change in quantity depending upon the situation.

Indirect proportion: Two variables are said to be directly proportional to each other if, whenever one changes by any factor, the other changes by the same factor. Also, the two variables are connected by an invariant rate.

Direct relationship: Whenever one variable gets bigger, the other does too. If one variable becomes smaller, so does the other.

Factor: A factor is any of the numbers, quantities, or symbols which, when multiplied together, form a product.

Indirect relationship: Same as inverse relationship.

Invariant rate: When all the members of a set of rates are equivalent to each other, they are invariant. Or, the rates that connect two variables that are directly proportional to each other are said to be invariant.

Inverse relationship: As one variable gets bigger, the other becomes smaller, and if the first variable becomes smaller, the other gets bigger.

Quantity: A quantity of anything tells how much there is, and is always expressed by a number and the unit of measurement.

Rate: The quantity of one variable per quantity of a second variable.

Rate proportion: An expression that equates two rates.

Ratio: The quotient of two numbers or of two quantities measured in the same units.

Ratio proportion: Two equivalent ratios for directly proportional variables arranged in an equality.

Reciprocal rate: A rate divided into the number one. Or, a rate inverted so that the denominator becomes the numerator, and the numerator becomes the denominator.

Relationship: Two variables have a relationship with each other if one changes whenever the other changes.

Unitary rate: The rate that tells how much of one variable per one quantity of the other variable.

Unit of measurement: The measurement label, such as pounds in 5 pounds or grams in 3½ grams.

ACTIVITY 1.1–The Pennies Activity (Guess the Number!)

Here is how to change a lucky guess to an educated estimate.

Purpose: On the teacher's desk, you will observe a container filled with pennies. Your problem is to find out how many pennies are in the container. Your purpose is to measure a rate and to apply the Unitary Rate Rule to solve a problem.

Equipment: You will be given a small pile of pennies, a round or square jar, and a ruler. The equipment will be collected at the end of the period.

Procedure: The class will be divided into teams of two students each. Teams may consult or observe other teams. Teams may also refuse to answer questions or to permit observers from other teams.

The task of the team is to make the most reasonable estimate of the solution to the assigned problem. A description of the procedure used to get the answer is required. You may not touch the container of pennies.

Clues: The teacher may offer one or more clues to the class if needed.

Report: Write a report showing your calculations on the number of pennies in the container. The report is to be neatly and carefully written in ink on lined 8½″ × 11″ paper using the format suggested by your teacher.

ACTIVITY 1.2–How Thick Is a Page?

Purpose: To measure the thickness of a page in this book.

Equipment: Several textbooks, centimeter ruler.

Procedure: How can you measure the thickness of a page in a book? That could be a difficult task with the equipment you have–unless you use your knowledge of rates. You will have to hold the pages of several books together to obtain precise data. Keep in mind that there are two numbered sides per one page, that you do not want to measure the thickness of the cover, and that some pages at the beginning and at the end of the book may not be numbered.

1. Measure the thickness of 60, 110, and 140 numbered sides. Record under (1) on the Report Sheet. Then, write the numerator over the denominator under "Rate" on the Report Sheet for each rate (rates for 30, 55, and 70 pages). Reduce each of these rates to the equivalent unitary rate (rate per one page) and enter on the Report Sheet. Be sure to include measurement units throughout.

   Next, add up all three unitary rates and divide by three to get the average thickness per page.

2. Answer (2). Note that this is not the rate calculated for (1), but can be calculated from it. (*Hint:* If there are 0.002 cm/page, then there is one page per 0.002 cm; 1 page/0.002 cm can be reduced to a unitary rate.)

3. On the front desk is a copy of the textbook with a bookmark in it. The uppermost numbered side is "Number 1." On what numbered side is the bookmark? Enter your data and calculations under (3).

4. Answer the remaining questions on the Report Sheet.

ACTIVITY 1.3–How to Count a Stack of Pennies

Purpose: To count pennies with a ruler.

Equipment: 15–40 pennies, centimeter ruler.

Discussion: Some banks count large amounts of pennies very simply. Tellers just stack them neatly in a round column and place them onto a slip of paper marked with the length of a column of fifty pennies. Then they wrap up the pennies in 50¢ packets. You may have seen cashiers at supermarkets break these open to place into the cash registers. The same can be done with nickels, dimes, and quarters. Larger coins are often hand-counted because an error means a bigger loss.

A. Obtaining the Data

Stack the pennies in a neat pile. Count how many pennies are in the pile and record in Answer (2) on the Report Sheet.

Measure the length of your stack of pennies and record in (2).

Remove two pennies from the pile. Write the new count of pennies and the new length in (3).

Remove two more pennies and repeat the measurements; record in (4).

B. Calculations

Calculate the average number of pennies in your measurements. Calculate the average length of the piles of pennies. Enter these in (5).

Calculate the unitary rate of pennies per centimeter and enter in (6).

Answer the remaining questions on the Report Sheet.

2

Equations with Meanings

2.1 INTRODUCTION

This chapter begins an exploration of natural relationships expressed through meaningful and useful shorthand. Instead of using words each time a variable is described, a system of shorthand will be used in which symbols are substituted for the names and quantities of variables. This will open up the entire world of algebra for use in problem solving in the sciences. Before this chapter is finished, you will be able to understand where the equations came from that you found in textbooks and, in the next chapter, you will even write some of these equations yourself.

The applications of algebra to problems of the home, business, and industry can be an invaluable aid to problem solving in these areas. In the natural sciences, too, mathematics helps in discovering that we do not live in a capricious disorderly world where strange forces change the rules before we have learned them. Rather, mathematics helps science to find *connections* that can be precisely described so as to make useful predictions.

Scientists have frequently described the awe with which they regard the majestic orderliness that can be found and used in nature. Mathematics is a splendid tool for this noble purpose.

2.2 IDENTIFICATION OF VARIABLES

The equations to be explored in this chapter will connect variables algebraically. The variables may be such things as length, volume, money earned, money spent, area, time, electrical charge, items purchased, and so on.

Before going on, it will be helpful to practice recognizing variables from their units of measurement. Variables that are met in everyday activities are usually easy to identify from their units, such as money whether expressed in francs, dollars, pesos or thousand-dollar bills. You may need a little help, however, with variables in the sciences. Following is a table of some variables encountered in science and some units that are used to measure them, both in science and in consumer usage. Those marked with an asterisk are the base units of the scientific International System of Units (*SI* units). Prefixes are used to make the base units larger or smaller as illustrated by some of the following units.

Name of Variable	Unit of Measurement	Abbreviation
length (or distance)	meter*	m*
	millimeter	mm
	centimeter	cm
	kilometer	km
	inch	in.
	yard	yd.
	mile	mi.
volume	cubic meter*	m^3 *
	cubic centimeter	cm^3
	millilitre	mL
	decilitre	dL
	litre	L
	gallon	gal.
	quart	qt.
	pint	pt.
	fluid ounce	fl. oz.
mass	gram	g
	milligram	mg
	kilogram*	kg*
time	second*	sec*
	minute	min.
	week	wk.
	year	yr.
weight	ounce	oz.
	pound	lb.
	ton	—
area	square centimeter	cm^2
	square meter	m^2
	square foot	$ft.^2$
	square yard	$yd.^2$
	acre	—

Identify the variable measured by each unit in the following:

1. cm.
2. pt.
3. hours.
4. millilitres.
5. cubic centimeters.

Some of the following can be identified not only by numerator and denominator variables but also by special names for the rate itself:

6. mi./hr.
7. $/item.
8. mL/g.
9. g/mL.
10. city block of land.
11. cubic miles.

Variables used in consumer and commercial activities may not always have such convenient terms for names. For example, the variables in the rate of 3 oz. wool per square foot are the weight of wool and the area. A broader term for weight of wool might be weight of fiber or maybe weight of material. Scientific variables are usually quite exact in their meanings whereas everyday terms may sometimes be less exact.

12. 10 mi./hr.
13. 10 g copper bar/cm copper bar.
14. 5 g salt/100 mL salt solution.
15. 100 pills aspirin/bottle.
16. 60 square feet/gallon paint.
17. There are 12 inches per foot.
18. There are 0.0833 feet per inch.
19. 100 Calories/apple.

2.3 REVIEW OF ALGEBRA

An equation you probably encountered frequently in algebra was $y = ax$. Usually, a was given some fixed value such as 5. Then the equation appeared as

$y = 5x$.

You may have been given some values for x and asked to calculate y. For example, if $x = 3$, y is calculated to be $5 \times 3 = 15$; if $x = 4$, $y = 5 \times 4 = 20$, and so on. Since the value of y depends on the value given to x, x in such an equation is called the independent variable and y the dependent variable. The value for a is called the fixed or constant value, or just the constant.

As a review, fill in the values for the dependent variable in the following:

1. Given $y = 5x$, calculate y when $x = \frac{1}{4}$.

2. Given $y = 310x$, calculate y when $x = 40$.

3. Given $y = \frac{x}{8}$, calculate y when $x = 40$. What is the constant in this equation?

So far, equations like $y = ax$ appear simply to be mathematical statements. When translated to the real world of measuring and counting, however, mathematics becomes a very powerful tool for calculations. It allows for concise and exact descriptions of ideas that might take many, many words to express. In fact, the words might take so many pages that we might never get past just trying to say them; we would never get to use or combine the ideas.

Starting with the abstract mathematical statement, $y = ax$, let us see what levels of relationships this expresses in the real, complex world in which we live.

2.4 THE EQUATION FOR VARIABLES RELATED BY A DIRECT PROPORTION

The first relationship to be studied algebraically here is the direct proportion. In this section, an equation will be presented that symbolically expresses a direct proportion. The sections following will explore the characteristics of this algebraic equation.

Recall that whenever two variables are directly proportional to each other, they are connected by an invariant rate. The idea of the invariant rate leads directly to a useful equation for any direct proportion, as we shall see. The equation is

$$\frac{y}{x} = a \qquad \text{or} \qquad y = ax$$

where y stands for the quantity of one variable, x for the quantity of a directly proportional variable, and a is an unchanging quantity called the constant or the rate constant.

In real life and in science, most numbers with which we work have units or labels. Thus, if we work with an equation in science such as $y = ax$, y does not stand for a number but for *a number of things*. Likewise, x stands for a number of other things. The symbols y and x are still called variables, but now we are talking of quantities, not of 5, 6, or 10 but of 5, 6, or 10 dollars or 19 students or 2000 hours.

Now, let us see how rates are connected to the equation $y = ax$, remembering that y and x are no longer abstract numbers but quantities of variables that can be counted or measured.

Another way of writing

$\frac{y}{x} = a$ is

$$a = \frac{\text{quantity of } y \text{ variable}}{\text{quantity of } x \text{ variable}}$$

where the $\frac{\text{quantity of } y \text{ variable}}{\text{quantity of } x \text{ variable}}$ can have many quotients, as long as they are all equivalent to a, and where a is some one convenient, or very familiar, rate such as the unitary rate.

To illustrate, the variables might be *money earned* by working and the *time* needed to earn it. The equation in symbols is $\frac{y}{x} = a$ where y is money earned, x is time, and a is a fixed or constant value of the rate. In words, this equation is $\frac{\text{quantity of money earned}}{\text{time}}$ = the constant rate. The quotient at the left may vary depending upon the amount of time that was worked, but it has to equal the constant at the right which is fixed at some selected value for the rate.

For example, suppose that the constant, a, is a salary of \$200/week. For time periods of, respectively, 9 weeks, 27 weeks, and 0.5 weeks, the following substitutions can be made into the equation:

$$\frac{y}{x} = a$$

$$\frac{\$1800}{9 \text{ weeks}} = \frac{\$200}{\text{week}}$$

$$\frac{\$5400}{27 \text{ weeks}} = \frac{\$200}{\text{week}}$$

$$\frac{\$100}{0.5 \text{ weeks}} = \frac{\$200}{\text{week}}.$$

We see that a equals all the various rates that can be reduced to \$200/week; it is the invariant rate, usually expressed as the unitary rate.

Converting these to the form $y = ax$, these become

$$y = a \quad x$$

$$\$1800 = \frac{\$200}{\text{week}} \times 9 \text{ weeks}$$

$$\$5400 = \frac{\$200}{\text{week}} \times 27 \text{ weeks}$$

$$\$100 = \frac{\$200}{\text{week}} \times 0.5 \text{ weeks.}$$

To summarize, all direct proportions can be described by the mathematical statement

$$y = ax$$

where y = quantity of one variable
x = quantity of another variable
a = fixed or constant rate of the y variable to the x variable.

The constant a is usually the unitary rate but this is not a rule. Whereas y and x may be any of a set of paired quantities, the value of a may not vary once it is established for the particular situation.

It is in no way necessary to use y and x in the equation for the direct proportion. In science, business, and industry, the variables are frequently indicated by the letter, sometimes capitalized, that begins the name of the variable, such as v for velocity, P for principal (in loans), P for population (in insurance), and so on. In the above equation, the total money earned could have been called M, total time period called t, and W used for wages/one time period. The equation then becomes $M/t = W$ or $M = Wt$. You can use any symbols you wish as long as they are defined, such as

$$\frac{\phi}{*} = ☺$$

where ϕ = total earnings
$*$ = time period
☺ = wage per each time period.

Most people prefer symbols that are easily available in printer's type, such as the alphabet. Also, symbols for which the name is known are preferred rather than those like ϕ (the Greek symbol, phi) and ☺.

$Ja = H$
where J = Jekyll
H = Hyde
a = constant rate.

The remainder of this chapter will be devoted to examining some of the important characteristics of the equation that can represent a direct proportion, $y = ax$. The following topics are to be considered:

1. solving the equation, $y = ax$, after substitution of actual quantities;
2. analyzing for consistency of units used in the equation;
3. checking that the equation satisfies the criteria for a direct proportion, and
4. analyzing any symbolic equation for variables that are directly proportional.

2.5 SOLVING EQUATIONS FOR A DIRECT PROPORTION

Given the equation $y = ax$, with two variables and a rate constant, if two variables are known, the third can be solved for algebraically. Three steps can be followed in solving such problems:

1. Identify the given equation and the meaning of each symbol.
2. Write down any other given quantities, and write what is wanted.
3. Place the quantities and rates into the equation and solve the equation algebraically.

Below is a sample problem for solving an equation for a direct proportion.

Sample Problem

Problem: Given that $M/t = W$ where M is money earned, t is the time period, and W is the weekly salary.

a. What is earned in $4\frac{3}{5}$ weeks at $200/week?

b. How many weeks are needed to earn $3100 at $225/week?

Solution: To solve this, first *identify the given equation and the meaning of each symbol.* In this problem, given

$M/t = W$

where M = money earned
t = time period
W = salary/week.

Secondly, *write down any other given, and write what is wanted.* We will do this for a:

$t = 4.6$ weeks, $M = ?$, and $W = \$200/\text{week}$.

Third, *the quantities are placed into the equation and the equation is solved algebraically* while maintaining the rules for manipulating units, as learned earlier.

$$\frac{M}{6 \text{ weeks}} = \frac{\$200}{\text{week}}.$$

(Multiply both sides by 6 weeks, or cross over the 6 weeks to the other side if that is what you learned.)

$$M = \frac{\$200}{\text{week}} \times 6 \text{ weeks} = \$1200.$$

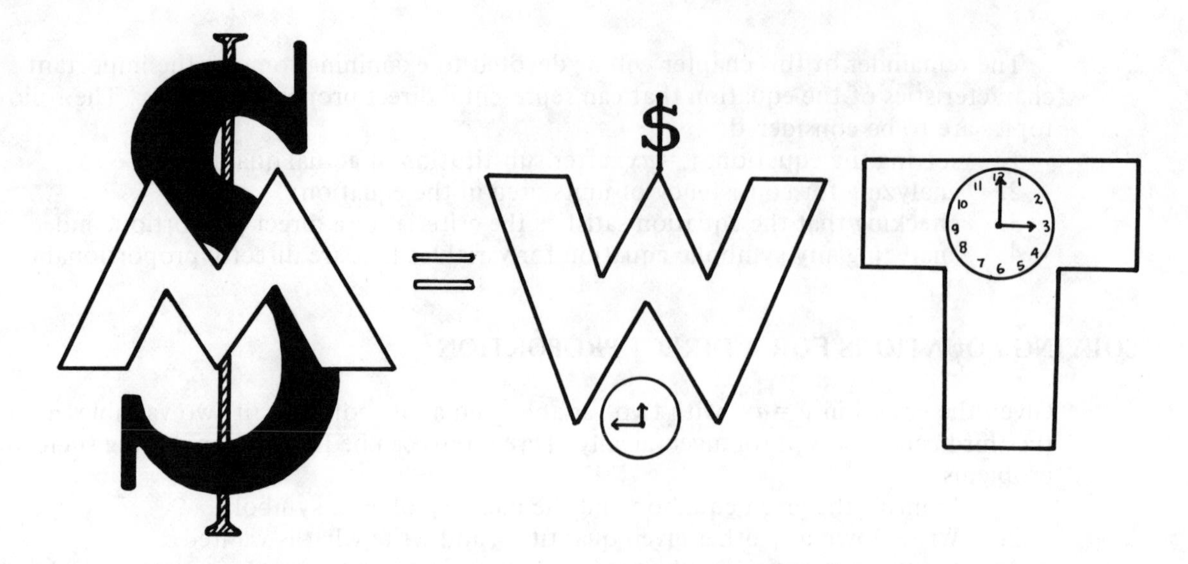

Now we will apply the second and third steps given above to part b. of the problem.

$M = \$3100$, $t = ?$, and $W = \$225/\text{week}$.

$$\frac{\$3100}{t} = \frac{\$225}{\text{week}}.$$

$$t = \frac{\$3100 \cdot \text{week}}{\$225} = 13.8 \text{ weeks}.$$

Apply the three steps toward solving the following problems algebraically.

1. Given the equation

   $C = PI$

   where C = total cost
   I = number of items
   P = price per item.

   a. Identify which are the two variables.

   b. If P is fixed at \$12/shirt and I = 8 shirts, what is C?

   c. If P is fixed at \$15/shirt,
   (1) what is the cost of 6 shirts?
   (2) what is the cost of 15 shirts?

   d. If C is \$120 and I is 8 boxes of copper sulfate, what is P?

   e. Given the value of P calculated from d., what is the cost of 3 boxes of copper sulfate?

   f. Given that C = \$40 and I = 6 pens, what is P?

   g. How can P be called a constant if it changed in each problem above?

2. Given that $D/T = S$ where D = distance covered, T = total time, and S = a constant which is distance/time:

   a. If S = 55 mi./hr. and T = 3.5 hrs., what is D?

   b. If D = 8 mi. and T is 2 hrs., what is S?

3. Use the equation $d = m/v$. In this equation, d stands for density, a constant rate. Density is the special name for the unitary rate of mass to volume of a substance; m/v stands for mass/volume.

   a. Use the table below to calculate the density of gold.

Mass of Gold (grams)	Volume of Gold (millilitres)
57.9	3
______	4
19.3	______
______	10

   b. Complete the table.

   c. Given that the density of mercury is 13.5 g/mL at room temperature. Use the density equation to calculate the volume occupied by 1.0 g of mercury.

4. Given that $B = aC$ where B = number of bottles, C = cost of bottles and a is the constant rate of bottles/cost, calculate with the aid of the equation the number of bottles that can be purchased for \$22 given that the constant rate is 3 bottles for \$5.

Notice that in Problem 4, a is a constant rate that is not a unitary rate. While the rate constant picked to represent the invariant rate is usually the unitary rate, there are cases where other types of constant rates are more familiar and so are used. Examples of such rates are 5 bottles/\$2.97 or one house/10,000 ft.^{2}. Any fixed rate of the y variable to the x variable may be used, as found convenient.

2.6 CONSISTENCY OF NUMBERS AND UNITS IN AN EQUATION

Whenever quantities are substituted into any symbolic equation, not only must the numbers be equal for both sides of the equation but the units of measurement must also be equal.

Let's take another look at the equation for the wage proportion from Section 2.5.

$M = Wt$

where M = total money earned
W = wages per fixed time period
t = total time.

Since W was given in units of dollars per week, M *has to be* in dollars and t *has to be* in weeks.

$$\text{dollars} = \frac{\text{dollars}}{\text{weeks}} \times \text{weeks.}$$

Had any other units been used for M and t, the units would not have been equal on both sides of the equation and so would have destroyed the equality.

The next question to consider is whether the units must always be in dollars and weeks. Suppose, for example, that the wages are paid monthly, that is, in dollars per month. The equation can still be used provided that t is also in months, not weeks.

In fact, W can be per second, minute, day, year, eon, or whatever, as long as t is measured in the same units. Moreover, the money can be measured in dollars, cents, yen, pesos, francs, megabucks, or beads as long as both M and W use the same money units.

Thus, an equation may be expressed symbolically or in words without specifying the units. The fact that the expression is an equality requires that *consistent* units be used, that is, that the units on one side of the equation must algebraically equal those on the other side of the equation. The rules for working with units learned in Chapter One apply likewise to equations.

As a result, equations have broader meanings than do invariant rates. Invariant rates are limited to the actual case they describe, such as \$18/hr. and rates equivalent to \$18/hr. Equations such as $M = Wt$ apply to any invariant rate in any units for those related variables.

For simple problems, units analysis may be used without bothering to work with an equation.* For less familiar variables, however, equations supply a very useful shorthand. It is much easier to work with an equation than with a long word statement about relationships. Moreover, equations permit working with more than two variables. Finally, there are many more relationships than the direct proportion, relationships that cannot be handled simply by units analysis.

Equations are used in many places in the sciences. The understanding of equations, therefore, is an invaluable key to the understanding of science.

To get back to the usage of units in equations, solve the following problems.

1. Given the equation $y = ax$, where y and x are variables and a is a constant, state which of the following violates the requirement for equality of units.

    a. a is measured in g/mL, y is measured in grams and x in millilitres.

    b. y is measured in centimeters, x in millilitres and a in grams/mL.

    c. y is measured in hours, x in miles and a in miles per hour.

    d. a is measured in kg/cm^3, y in cm^3 and x in kg.

    e. a is measured in pills/bottle, y in pills and x in bottles.

2. In the following, given the units for y and x in $y = ax$, state the units for a:

	x	y	a
a.	dollars	gold coins	____________

*The method of units analysis, also known as dimensional analysis, factor-label method, and the quantity calculus, is a special method of problem solving used in problems dealing with rates and is presented in Module One of this series.

	x	y	a
b.	calories	grams	______________
c.	pins	box	______________
d.	gold coins	dollars	______________

3. For each of the following, fill in the missing units and identify which is y, x, and a in the equation $y = ax$.

   a. $\square = \frac{\text{litres}}{\text{quarts}} \times \text{quarts}.$

   b. $\text{cm} = \frac{\square}{\square} \times \text{sec}.$

   c. $\square = \frac{\text{cans}}{\square} \times \text{cases}.$

4. From the arrangement of the units for D, T, and S in Problem 2a., Section 2.5, show that $D = TS$ and that D does not equal $\frac{S}{T}$.

2.7 THE EQUATIONS AND CRITERIA FOR A DIRECT PROPORTION

The two criteria for deciding whether or not two variables are related by a direct proportion are:

1. *The Invariant Rate:* The quotient of all paired quantities of the variable is invariant.
2. *Constant Factor:* When one variable changes, the other changes by the same factor. That is, the ratio of the new to the former quantity for one variable equals the ratio of new to former quantity for the other variable.

Either of these criteria may be used to determine if proportionality exists.

The algebraic equation $y = ax$ must of itself meet both of these criteria if it does, indeed, describe a direct proportion.

We have already recognized that $y = ax$ meets the requirement of the invariant rate since y/x is the quotient of two variables that equals the constant a.

Let us next examine the second requirement.

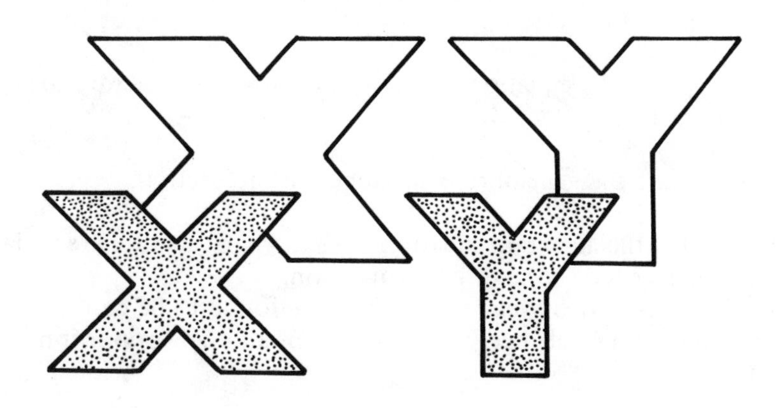

First, let a be a constant; we will deal only with changes in the variables y and x and not allow a to change. Suppose that x changes to a number five times as big as it was before. What must happen to y in the equation $y = ax$?

What is your answer?

If x becomes $5x$, then since $(?)y = (a)(5x)$, y will have to change to maintain the equality; a is not allowed to change. Therefore

$$5y = a \cdot 5x.$$

We see that y has changed to $5y$. That is, when x becomes $5x$ (five times as big), y changes to $5y$ (also five times as big).

Another way of seeing this is to give a some invented value and then see what happens to y for various values of x. For example, let $a = 3$. Then

$$y = 3x.$$

If $x = 1$, $y = 3$.
If $x = 5$, $y = 15$.

We see that when x changes from one by a factor of 5 to become 5, y changes from 3 to 15 or also changes by a factor of 5, as was also shown by the other method.

Answer the following, given that $y = ax$ where a is constant.

1. When x changes by a factor of 8, what happens to y?

2. When x changes by a factor of $\frac{1}{10}$, what happens to y?

3. When x changes by a factor of 0.02, what happens to y?

4. When y changes by a factor of 4, what happens to x?

5. When x changes by any factor in the equation $y = ax$, what happens to y?

6. Given $q = rs$ where s is constant. If r changes from 10 plops to 50 plops, what happens to q?

7. Given $q = rs$ where r is constant. If q changes from 10 bananas to 1 banana, what happens to s?

8. Given the equation $y = 6x + 2$, if x changes by a factor of 3, does y change by the same factor?

9. Given the equation $yx = a$, if x changes by a factor, does y change by the same factor?

10. State the general rule developed in this section.

We see that the algebraic equation $y = ax$, where y and x are variables and a is constant, satisfies both criteria for a direct proportion.

It follows that whenever two variables are found to be connected by an invariant rate or by the same ratio of change, the equation for a direct proportion, $y = ax$, may be written for them.

2.8 READING EQUATIONS TO IDENTIFY PROPORTIONS

A. Equations with Two Variables

Not all equations in science and industry (or psychology, nursing, business, or shop), look like $y = ax$, although many of the ideas of these disciplines will be presented to you in the form of an equation.

Every equation in science deals with natural relationships—with the way variables are related. As we have mentioned, this is very different from pure mathematics, which handles only the way symbols (without regard to any real meanings) can be manipulated through addition, multiplication, and so on. A scientific equation shows real relationships, some simple, some complex, but all about things that can actually be measured. An equation cannot be developed to describe real things, cannot be set up, without first taking measurements to find out how the variables are related. After the scientist discovers that the tables of data show, for example, a direct proportion, then the equation is set up. This equation is later what is presented to the reader in a textbook.

We will first look at some equations to see if any of the variables are related by a direct proportion. Then in the following chapter we will set up some equations from tables of scientific data.

In the following problems, state whether the two variables specified are proportional. In each equation, k is a constant.

1. Distance, d, that a car travels in time, t, where $d = kt$.

2. Distance, d, that an object falls in time, t, where $d = kt^2$.

3. Concentration of solid, C, dissolved in volume of solution, V, where $CV = k$.

4. Weight of solid, W, dissolved in volume of solution, V, where $W = kV$.

5. Degrees Celsius, °C, per degrees Fahrenheit, °F, where $°C = \frac{5}{9}(°F - 32°)$.

6. People standing, P, per people sitting, p, in a room with 40 seats, where $P + p = 40$.

B. Equations with More Than Two Variables

Some equations in the natural sciences can look very complicated, yet it is quite easy to pick out which variables are directly proportional. Simply hold all the other variables constant except the two you are examining and then you can make your decision.

For example, are F and a directly proportional in $F = ma$? To find out, hold m constant. The answer is yes for the following reasons, either or both of which may be used. By holding m constant, we can say that $F = ka$ where k stands for a constant; therefore, (1) whenever a changes by any factor, F changes by the same factor and (2) the rate, F/a, is invariant (is constant) since it equals k. However, m and a are not proportional in this equation for the following two reasons. By holding F constant, we can say that $ma = k$ where k stands for the constant value; then, (1) when m changes by any factor, a does not change by the same factor (a gets smaller when F gets bigger, and a becomes bigger when F becomes smaller), and (2) m/a is not an invariant (or constant) rate (it is m times a that is constant).

As another example, consider the equation $y/z = q$. Are any of the variables directly proportional? Let us first consider y and q. Holding z constant (equal to k), we can say that $y/k = q$ or that $y = kq$; hence, y and q change by the same factor. We can also say that the quotient $y/q = k$ so the rate of y to q is invariant. We conclude that y and q are proportional. Likewise, y and z are proportional. However, z and q are not proportional; z and q do not change by the same factor since $qz = k$ (q gets smaller when z gets bigger and q gets bigger if z becomes smaller), and z/q is not an invariant rate.

However, it is not necessary to make such an analysis each time to decide on your answer to the above; a short-cut can be used instead. All that is needed is to check whether the two variables are related by the same form of equation as $y = ax$ (or $y/x = a$). If the symbols for the two variables are in the same y to x arrangement, then the two variables are directly proportional.

Let us see if x and y are directly proportional in the following sample problems.

Sample Problem One

Problem: Are x and y directly proportional in the equation $y = kx$?

Solution: Since y and x are in the same arrangement in $y = kx$ as in $y = ax$, they are proportional.

Sample Problem Two

Problem: Are y and x directly proportional in $yx = k$?

Solution: There is no way to rearrange $yx = k$ algebraically to $y = kx$. The variables are not directly proportional.

Sample Problem Three

Problem: Are x and y directly proportional in the equation $y = zn^2x$?

Solution: To solve this, first think of zn^2 as a constant and equal to a, so $y = ax$. From this, we can see that y and x are directly proportional.

Sample Problem Four

Problem: Are y and x directly proportional in the equation $zq = yxb$?

Solution: To solve this, first think of zq/b as unchanging and equal to some constant such as a, so $a = yx$. There is no way to rearrange this to $y = ax$. Hence, y and x are not directly proportional.

In the equations following, state whether y and x are related by a direct proportion.

7. $\dfrac{y}{x} = c.$

8. $x = yz.$

9. $y = bx.$

10. $ay = cx.$ (*Hint:* If a and c are both constants, then c/a is also a constant, and $y = \frac{c}{a} \cdot x.$)

11. $y = ax^2.$

12. $y = \frac{rx}{m}.$

13. $y^2 = ax.$

14. $y = ax + b.$

15. $y = \frac{ax}{mn}.$

16. $yx = a.$

17. $\frac{ay}{b} = \frac{rx}{t}.$

18. $y = \frac{m^2 x}{r}.$

C. If x Changes, What Happens to y?

We saw in Section 2.7 that, given $y = ax$ where a is held constant, whenever x changes by any factor, y changes by the same factor, and vice versa. This gives a way of predicting what happens to one variable when another changes, even though the quantities may not be known for the two variables and even though the rate constant may be unknown.

19. If y becomes 30 times as big, by what factor must x be multiplied?

20. If y is multiplied by the factor of z, what must happen to x?

21. If y changes from 8 to 16, what happens to x?

22. If y changes from 40 to 10, what happens to x?

23. If y changes from ½ to 1, what happens to x?

24. Given the equation, $y = \frac{ax}{t}$, how does y change when x triples, all else being held equal (that is, unchanged)?

25. Given the equation $y^2 = ax$, how does y^2 change when x changes?

26. Given $y = ax^2$, what changes by the same factor as y does when a is held constant?

27. Given $y = zx$, if both z and x may change, what must double when y doubles?

2.9 LAST WORDS

This chapter introduced the idea of using a mathematical equation to show the relationship between two real-world variables. The equation for the variables symbolized by y and x when they are proportional to each other is given by $y = ax$ where a is a constant which is the fixed rate for the two variables. Usually, a is a unitary rate.

Mathematical equations for real-world variables follow the same laws of mathematics as do algebraic symbolic equations. The equations must also show an equality of units on both sides of the equation.

The equation that shows a direct proportion between two variables, $y = ax$ or the rearranged form $y/x = a$, meets the two criteria for a direct proportion both algebraically and with real-world variables. These requirements are: the invariant rate for y/x, and the same ratio of change for both variables.

By comparing equations to the form $y = ax$, it can be determined whether or not two of the variables in the equation are directly proportional. Whenever two variables in an equation are found to be directly proportional, then the change in one variable can be predicted from the change in the other variable.

DEFINITIONS

Constant rate: A given member of a set of equivalent rates.

Dependent variable: The variable in an equation whose quantity is to be calculated.

Fixed rate: Same as constant rate.

Independent variable: The variable in an equation to which has been assigned a given quantity.

Rate constant: Same as constant rate.

ACTIVITY 2.1–The Turn of the Screw

Purpose: To measure length using a C-clamp as a measuring tool.

Equipment: One C-clamp, cm ruler.

Procedure:

1. Turn the handle of the C-clamp and observe what controls the amount of downward and upward motion of the screw. Count and record the number of threads per cm. Raise the screw one cm from the base while counting the number of turns of the handle required for this. Repeat for 2 cm and 3 cm (or 1½ cm and 2 cm for a small clamp). Answer questions (1) to (7) on the Report Sheet.

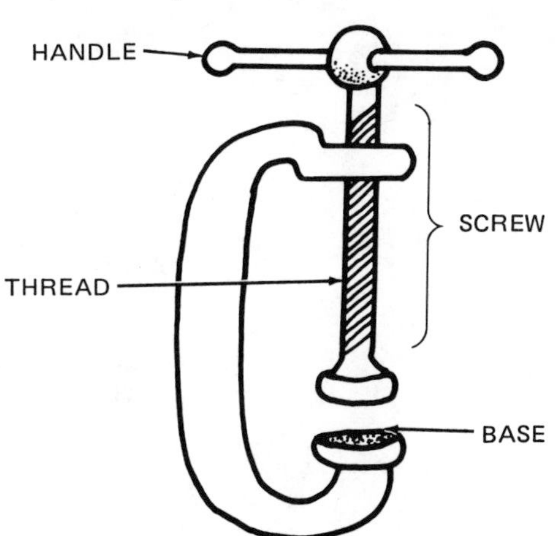

2. Your C-clamp can be used to measure lengths of irregularly shaped objects and of narrow widths. Your teacher will give you two objects to measure. Use only your C-clamp; do not use the ruler! Each time a measurement is taken, first back the screw off with some turns, insert the object, and turn the screw until it is tight. Answer (8). After completing Chapter 4, answer (9) to (11).
   A device called a micrometer is used in the same way as the C-clamp, except that it has a collar on it that gives the reading without need for calculations.

3

How to Write an Equation for a Direct Proportion

3.1 EQUATIONS BY INSPECTION

Now that you have learned to "read" the equation for a direct proportion, you are ready to learn how to write one.

This chapter will show you some steps that you can regularly follow to construct such equations easily. Sometimes, instead of following such rules, equations can be developed just by looking at the information available. Before going on to examine rules for writing equations, suppose you try to do two problems by inspection. These problems are as follows:*

1. Write an equation that uses the variables S and T to represent the following statement: "There are sixteen times as many students as teachers at this school."
   Use S for the number of students and T for the number of teachers.

2. Write an equation that uses the symbols F and H to represent the following statement: "At the school cafeteria, for every four students who ordered frankfurters, five ordered hamburgers."
   Let F represent the number of frankfurters ordered and let H represent the number of hamburgers ordered.

*Similar problems were used in the Cognitive Development Program at the University of Massachusetts, Amherst, MA and reported in a paper entitled "The Role of Preconceptions and Representational Transformations in Understanding Physics and Mathematics," by Frederick W. Byron, Jr. and Jack Lockhead, September, 1979.

First write the equations for these problems, then continue reading to find the answers. The answer to Problem 1 (omitting any measurement units) is:

$S = 16T.$

The answer to Problem 2 (also omitting the measurement units) is:

$5F = 4H$ or $F = \frac{4}{5}H$

If you wrote $16S = T$, this is incorrect since, if there are 16 students, $16S = 16 \times 16 = 256$, or 256 teachers for 16 students—obviously wrong. The equation $16S = T$ says that the number of teachers equals 16 times the number of students, whereas it is really the number of students that equals sixteen times the number of teachers.

You may be interested to learn that 37% of a group of college engineering students incorrectly solved a problem like Problem 1 while 73% incorrectly solved one like Problem 2. They didn't have the advantage of studying a book like this one.

Keep in mind that an equation sets up both sides to be equal. It is necessary to decide in Problem 1 what you need to multiply T by to have the number equal S. The same idea applies to Problem 2, a more difficult problem where you need to find the number needed to multiply the number of frankfurters to have it equal the number of hamburgers.

Try another problem.

3. Given that there are 100 calories to every apple. Write an equation to represent this statement using C for the number of calories and A for the number of apples.

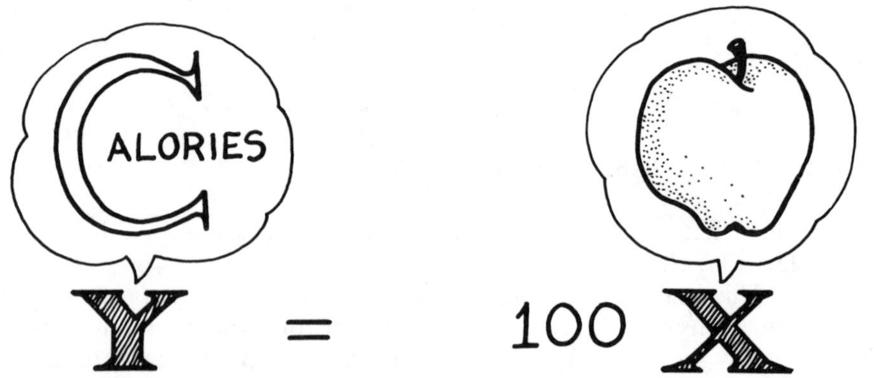

Notice that each of the equations for Problems 1-3 is of the form $y = ax$, the equation for a direct proportion. In *each* case, y and x are variables while a is a constant number.

Equations such as $S = 16T$ and $F = \frac{4}{5}H$ are correctly used as written only when units are omitted. For example, given $S = 16T$ and given 5 teachers and omitting units, $S = 16 \times 5 = 80$ so the number of students equals 80. Unit-free calculations such as this are common in algebra courses.

If, however, units are used for S and T, then units must also be used for the constant to maintain the equality.

80 students = 16 ? × 5 teachers.

Analysis of the units shows us that

$$80 \text{ students} = 16 \frac{\text{students}}{\text{teacher}} \times 5 \text{ teachers.}$$

Evidently, the constant is a rate, the rate of students per teacher. This agrees with what was presented in Chapter Two. It also agrees with the statement made earlier in this section that it is necessary to decide in Problem 1 what you need to multiply T by to have the number equal S. The number needed to multiply T, quantity of teachers, to get S, quantity of students, is the rate of students per teacher.

As we shall see in the next section, it is advantageous in most equations in science to use units. Hence, the constant in scientific equations is given as a rate rather then as just a number.

One more point to consider is whether it is even correct to use real-world equations without units. Since the measurement units on one side must equal the measurement units on the other side of any equation, they can all be cancelled out together, so there is nothing wrong with doing it.

$$16 \frac{\text{students}}{\text{teacher}} \times 5 \text{ teachers} = 80 \text{ students.}$$

$$16 \qquad \times 5 \qquad = 80.$$

Yet, cancelling the units destroys all the advantages of analysis of units. In those parts of algebra that deal with abstract mathematics, numbers without units must be used. However, when mathematics is used for real-life problem solving, the use of measurement labels can be very helpful and advantageous.

In the next sections, we shall consider some simple rules for writing the equation for a direct proportion, which will help to avoid setting up wrong equations such as $16T = S$ and make it possible to easily arrive at equations such as $F = \frac{4}{5}H$.

3.2 RULES FOR CONSTRUCTING EQUATIONS FOR A DIRECT PROPORTION OF THE FORM $y = ax$

A. How to Know that the Variables are Directly Proportional

To write an equation, it is necessary to know in advance how the variables are related. If they are connected by a direct proportion, the problem may show this in three ways:

1. The information that the variables are directly proportional is stated in words.
2. The invariant rate is given.
3. Data on pairs of quantities for the two variables is given. The data is examined to see if it satisfies either or both of the two criteria for a direct proportion, namely, the invariant rate and the same factor of change.

The last way is the one used in science to analyze relationships between variables. In scientific experiments, measurements are taken of at least several different quantities of one variable and of the corresponding quantities of the other variable. From this, it can be determined if the variables are directly related and, if so, whether they are also directly proportional.

B. The Rules

The rules that may be used for writing the equation, once it has been decided that the variables are directly proportional, are not complicated. They are as follows.

THE RULES FOR CONSTRUCTING THE EQUATION

1. ***Select and identify symbols for the variables.***
Decide on a name for each variable and select a symbol for each. Write these down.

2. ***Identify the invariant rate (or rate constant).***
Select a symbol for the rate constant and then define it in words.

3. ***Write the equation.***
Write the equation in the form y = ax *or* y/x = a. *The numerator variable of the rate constant goes where* y *is. The denominator variable of the rate constant goes where* x *is. The rate constant goes where* a *is in* y = ax *or* y/x = a.

4. ***Check the equation.***
Check that the units or names of all of the symbols in the equation satisfy the equality.

In the next three sections, these rules will be illustrated for problems that supply each of the three kinds of information that may be given as listed in Section 3.2A.

3.3 THE EQUATION, GIVEN THAT THE VARIABLES ARE DIRECTLY PROPORTIONAL TO EACH OTHER

Sample Problem One

Problem: Given that the mass and volume of a liquid substance are directly proportional to each other, write an equation to show this.

Solution: Rule 1. Select and identify symbols for variables:

Let m = mass,
v = volume.

Rule 2. Identify the rate constant:

The constant rate could be mass per volume or volume per mass. Without inside information, it doesn't matter which way you do it as long as you identify it. Scientists often use mass/volume, so let d = mass/volume (constant rate).

Rule 3. Write the equation:

$m = dv$.

Rule 4. Check the equality:

$$\text{mass} = \frac{\text{mass}}{\text{volume}} \times \text{volume}.$$

It checks.

Suppose that the invariant rate v/m were used instead of m/v. A different symbol would have to be picked for v/m since it is not the same as the defined d. Let q = volume/mass. The new equation is $v = qm$ (not $m = qv$); checking,

$$\text{volume} = \frac{\text{volume}}{\text{mass}} \times \text{mass}.$$

Sample Problem Two

Problem: Given that the total volume of an oil is directly proportional to the number of jars used. Write the equation.

Solution: (Select symbols):

Let v = volume of oil,
n = number of jars.

(Identify rate constant): r = volume oil/jar.

(Write equation): $v = rn$.

(Check): $\text{volume} = \frac{\text{volume}}{\text{jar}} \times \text{jar}.$

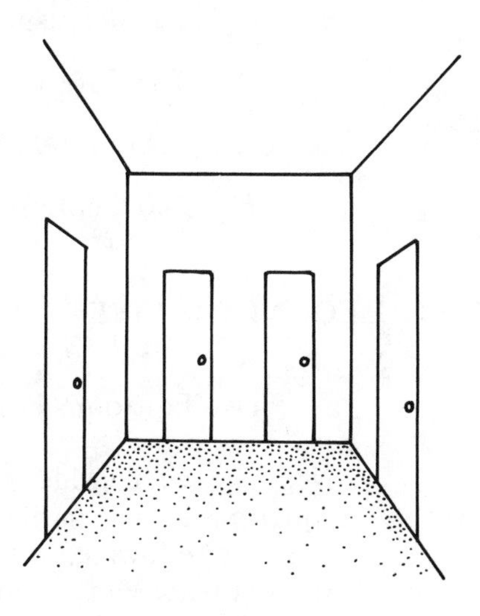

Now you are ready to write some equations!

Write general equations, identifying the symbols you select, for each of the following. All expressions are for variables related by a direct proportion.

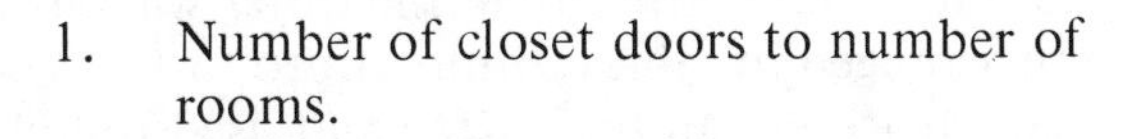

1. Number of closet doors to number of rooms.
2. Number of pages per this month's copy of a magazine.
3. Number of innings and number of games.
4. Volume and pressure of helium gas.
5. Volume of paint and area to be painted.
6. Quantity of staples and quantity of boxes.
7. Mass and volume of oxygen.
8. Height of a snowfall per height of rain water.

3.4 THE EQUATION, GIVEN THE INVARIANT RATE

An equation may be written when a rate is identified as invariant. In the following, rate constants are given. Write an equation for each to give as broad a meaning as you can to each variable. For example, suppose the given rate is 5 g linseed oil/cm^2 surface. The variables could be mass of linseed oil and the area of surface or mass of oil per area of surface or even mass of liquid per area of surface. Any of these is correct although the last one is the broadest.

1. 10 mi./hr.
2. 10 g copper bar/cm copper bar.
3. 5 g salt/100 mL salt solution.
4. 100 pills aspirin/bottle.
5. There are 12 inches per foot.
6. There are 0.0833 feet per inch.
7. 100 Calories/apple.

3.5 THE EQUATION, GIVEN DATA

A. Two Equations Possible per Set of Data

Next, we will consider the case of writing an equation from data. As was stated earlier, equations are ordinarily developed out of data rather than the other way around.

In the following, sets of data are arranged in tables. If direct proportions can be shown to exist between the variables, then equations can be written and used following the rules as applied to earlier sections.

Since every rate can be expressed in two ways, one of which is the reciprocal of the other, *there are two possible ways to write up the equation for any direct proportion* in the forms of $y = ax$ or $x = ky$. Following is a sample problem showing this.

Sample Problem: Given the table below, determine if it is a direct proportion. If it is, write the two equations that fit the data.

Energy (Calories)	Time Cycling (hours)
400	1
200	½
1200	3

We see that this is a direct proportion since each rate reduces to 400 Calories per hour. Another proof is that if the time halves or triples, so does the energy. The two equations are:

$E = kt$ where E = energy expended
t = time of cycling
k = energy/time
= 400 Cal./hr.

and $t = aE$ where a = time/energy
= 0.0025 hr./Cal.

Notice that for the equation $E = kt$ or $E/t = k$, t is the independent variable (see Section 2.3) and E is the dependent variable. For the equation $t = aE$ or $t/E = a$, E is the independent variable and t is the dependent variable.

Calculate the following.

1. For the sample problem above, given that k = 400 Cal./hr., a = 0.0025 hr./Cal., and that the bicycle is used for 48 hours:

   a. Use $E = kt$ to calculate E.

   b. Use $t = aE$ to calculate E.

2. Suppose that the rate constant, k, for the sample problem equals 300 Cal./hr. What is the value of a expressed as a unitary rate?

B. Given Data, Write the Equation

For each of the following problems, first decide whether the variables are related by a direct proportion. If so, write an equation, specify the value for the rate constant, and use your equation wherever calculations are needed. Since every rate relationship can be expressed in two ways, one of which is the reciprocal of the other, two equations can be written for each problem, as shown in the preceding Section A. Although the symbols for the two variables will be the same in each equation, the rates and the arrangement of the symbols must differ. You may select any symbol you wish for each rate constant, but it is customary to use either k or a.

3.

Soap Powder (lbs.)	Cost ($)
1.2	.59
3.6	1.65
4.8	2.05

4.

Salad Dressing (mL)	Oil (g)
12	3
16	4
24	6
28	7

5.

Area (feet squared)	Paint (gallons)
1020	2
2550	5
102	0.2

a. Write equations as per instructions.

b. Use one equation to calculate paint needed for 1000 square feet of area.

6.

Litres	Quarts
1	1.06
4	4.24
5	5.30
9	9.54

a. Write equations per instructions.

b. Use your equations to calculate:

(1) How many quarts are in 5½ litres.
(2) How many litres there are per 5½ quarts.

7.

Square Yards	Square Feet
2.5	22.5
7.1	63.9

a. Write equations as per instructions.

b. Use your equations to solve the following:

(1) Convert 369 square feet of room size to square yards.
(2) Convert 30 square yards of tile to square feet.

8. The longer a vat of chemicals is mixed for a certain reaction, the longer it takes to settle.

Settling (hrs.)	Mixing (hrs.)
2	14
½	3½
20	140
6	42

a. Write equations as per instructions.

b. If the mixture is stirred for 7½ hours, how long should the workers allow it to settle?

9.

Grams Fudge	Calories
50	150
150	450
300	900
500	1500

10. Let us take another look at Problem 9. An appropriate equation is $C = aM$ where M = mass fudge, C = energy in Calories and a is a constant whose value is

$$\frac{150 \text{ Cal.}}{50 \text{ g fudge}} \quad \text{or 3 Cal./g fudge.}$$

   a. What is the value of the rate constant in Cal./lb. given that 1 lb. = 454 g? What will be the units for C and M?

   b. What is the value of the rate constant in Cal./kg given that 1000 g = 1 kg? What will be the units for C and M?

We see that the equation may be used for various units as long as they are appropriate and consistent.

11. Given the equation $W = St$ where W = wages earned, S = salary, t = time:

   a. What are the units for W and t if S is in francs/day? In dollars/year?

   b. Do the number and unit for S change when the monetary system changes?

   c. What is meant by saying the units in any equation should be consistent and appropriate?

12.

Log of Equilibrium Constant (unit-free)	Voltage (volts)
1.00	34
0.80	27
0.60	20
0.05	1.7

Let $\log_{10} k$ = log of equilibrium constant,
V = voltage.

   a. Write equations as per instructions.

   b. Given that the voltage is 25 volts, what is the log of the equilibrium constant?

3.6 HOW A SCIENTIST DEVELOPS EQUATIONS

Now that you have studied how rates are related to equations, you are ready to carry out a few of the same kinds of derivations that a scientist does. Try the following.

1. A physicist times a car moving along a circular track to find out the speed of the car. The car moves at a constant speed. It covers two miles in 9 seconds, four miles in 18 seconds, and six miles in 27 seconds. Set up a rate table and an equation. (You must always, of course, define the symbols that you use when you set up the equation.)

2. A chemist gathers data on the mass of silver and the volume it occupies at room temperature. The following table of data is collected.

Mass (g)	Volume (mL)
1.0	.095
1.8	.17
2.4	.23

Set up an equation showing the relationship between mass and volume of a solid such as silver.

There is a special name for the unitary rate when it is mass/volume. What is it? What is its value for silver?

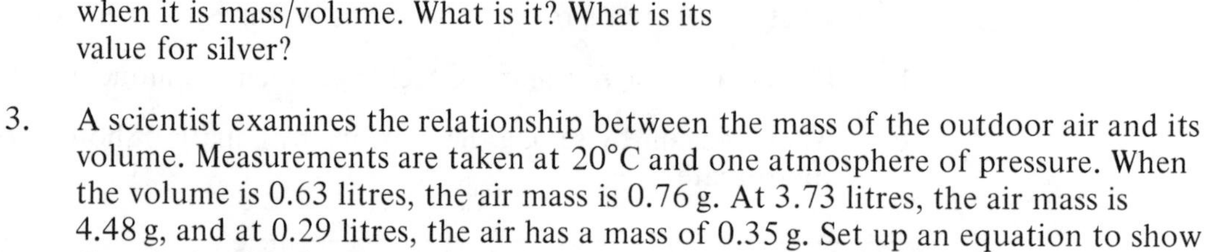

3. A scientist examines the relationship between the mass of the outdoor air and its volume. Measurements are taken at 20°C and one atmosphere of pressure. When the volume is 0.63 litres, the air mass is 0.76 g. At 3.73 litres, the air mass is 4.48 g, and at 0.29 litres, the air has a mass of 0.35 g. Set up an equation to show the relationship between air mass and air volume.

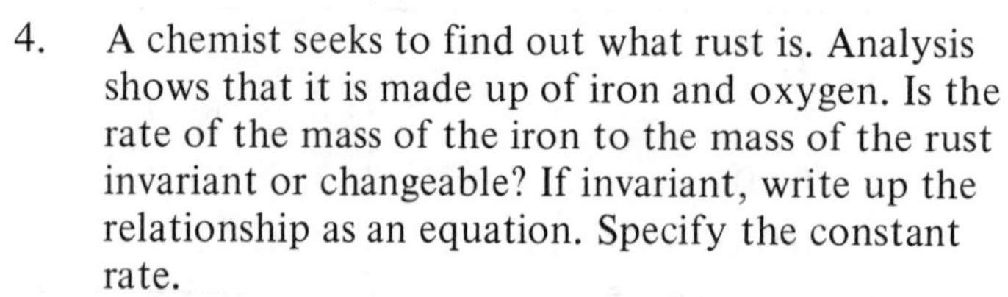

4. A chemist seeks to find out what rust is. Analysis shows that it is made up of iron and oxygen. Is the rate of the mass of the iron to the mass of the rust invariant or changeable? If invariant, write up the relationship as an equation. Specify the constant rate.

Use the equation to calculate the quantity of iron in 450 lbs. iron rust.

Mass Rust Analyzed (g)	Mass Iron (g)	Mass Oxygen (g)
50	35	15
100	70	30
135	94½	40½
150	105	45

5. This problem is tougher.

A chemist seeks to find out how many grams of oxygen react with one gram of hydrogen. Various combinations of hydrogen gas and oxygen gas are mixed and then sparked with an electric short-circuit. The reaction is explosive (as in the Hindenburg Zeppelin disaster), and water is formed. The left-over (final) gas is weighed and analyzed and the following information is obtained.

Initial Hydrogen	Initial Oxygen	Final Hydrogen	Final Oxygen
10 g	10.0 g	8.75 g	0
1 g	10.0 g	0	2 g
10 g	1.6 g	9.8 g	0

Is the mass of hydrogen used proportional to the mass of oxygen used? If so, then set up the equation.

3.7 CLASSY RATES

Equations help us to see that certain rates have special meaning because they describe some property that is well known. For example, when the rate of distance traveled per time is constant, the constant has a special meaning.

$$a = \frac{D}{T}$$

where D = distance traveled,
T = time.

The constant a in this case is commonly called the speed.

Can you think of any other rates that are in a class by themselves?

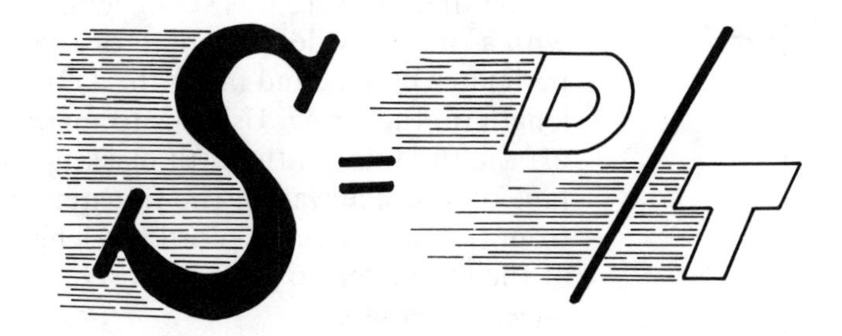

3.8 LAST WORDS

Equations in science are always meant to describe real relationships. Much of the time, they are based on relationships discovered by measuring real quantities. Other times, they describe relationships that are believed to exist but have not yet been supported by much evidence, if any. This latter is the supreme detective work of the scientist, an exercise of the imagination into what possibly might be in our universe. Equations are not confined to science. They are used by accountants, banks, surveyors, carpenters, machinists, electricians, and others.

In this chapter, a mathematical expression was constructed showing the relationship between two variables that are proportional to each other. The relationship is described by the equation $y = ax$, where y and x are variables and a is the constant rate. You do not need to be a professional mathematician to construct such equations. This chapter showed strategies and procedures for students in this class to write up such equations.

DEFINITION

Rate constant: Another definition for the rate constant, based upon Chapter Three, is that it is the constant k in $y = kx$, where y and x are variables. It is also sometimes called the constant of proportionality.

ACTIVITY 3.1—Bicycle Pedaling

Purpose: How does a ten-speed bicycle or a three-speed bicycle work? To find out, this activity will examine the ratio of turns of the rear wheel to turns of the foot pedal.

Equipment: Three- or ten-speed bicycle, masking tape, string, ruler.

Procedure:

1. Place a thin strip of masking tape along the outside wall of the rear tire across the width in line with the valve stem. Lay a piece of string over the outside rim of the tire so that it equals the perimeter of the tire. Measure this string length with a ruler. Divide the length of the perimeter by ten, and mark the 1/10 length on the string. Use this to mark off the tire into tenths with masking tape on the side wall of the tire starting from the valve stem. Record the length of the perimeter on your Report Sheet under (1).

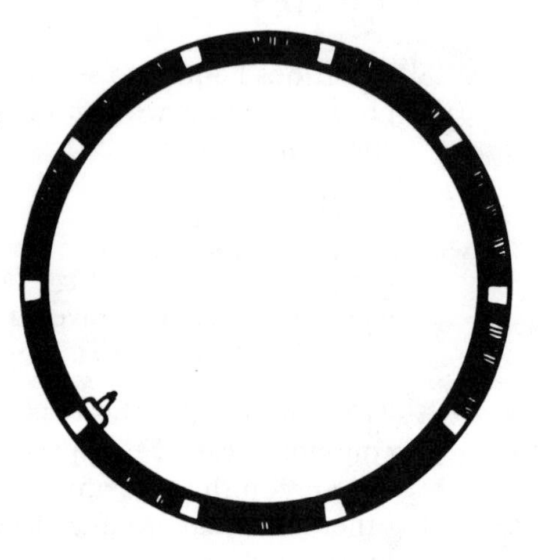

2. Set the bicycle gear to the lowest gear setting (for uphill). Turn the bicycle carefully upside down. Your task is to find the number of turns of the rear wheel per turn of the foot pedal. Try turning the foot pedal with one hand to see how to

stop it from free-wheeling when the foot pedal is set at vertical. Then practice turning the rear wheel several times with one partner braking the wheel with one hand just enough to prevent the wheel from free-wheeling. Next, set the foot pedal vertically and the rear wheel with the valve stem at the bottom. Turn the foot pedal slowly while the rear wheel is allowed to turn without free-wheeling, and count the number of turns of the rear wheel for one complete turn of the foot pedal. Estimate to 1/10 of a turn. Count turns for lowest speed, middle, and highest speed gears on the bicycle, and enter data on the chart in (2) on the Report Sheet.

3. Answer questions (3) to (8).

Your teacher may ask you to gether data now for the experiment entitled "Bicycle Gear System." In that case, you will apply the data when you get to that experiment and will make a separate Report Sheet to keep until then.

4

Inverse Proportions

the constant product or seesaw relationship

This chapter will continue our exploration of algebraic ways of stating real relationships in our world by investigating a relationship we have not yet studied, the inverse proportion.

4.1 EQUATION FOR THE SEESAW RELATIONSHIP

We have seen that there are a great many relationships in our world that are direct proportions. Equations of the form $y = ax$ show the relationship symbolically. Are there other relationships that can be described as simply, just using y, x and some constant a?

The answer is yes; there are two other such relationships. One is the additive-subtractive relationship, illustrated by a statement such as "Enid is two years older than José." The equation is $y = x + 2$, where y = age of Enid in years and x = age of José in years. You already know this type of relationship well, and we need not go into it here.

The other relationship is a special type of the inverse relationship we have already examined briefly. In an inverse relationship, as one variable gets bigger, the other gets smaller. For example, given $50 to spend, the *more* expensive the items purchased, the *fewer* can be bought.

A special type of inverse relationship is the inverse proportion. *In an inverse proportion, as one variable gets bigger, the other gets correspondingly smaller so that the product of the two variables remains constant.*

The equation that expresses this is

$$yx = k.$$

The way one variable goes up as the other goes down in illustrated below for the constant product 8. We will select values of y equal to 2, 4, 8 and 40.

y	•	x	=	k
2	×	4	=	8
4	×	2	=	8
8	×	1	=	8
40	×	$\frac{1}{5}$	=	8

The direct and inverse proportions have been compared to an elevator and a seesaw. In an elevator, everything goes up or goes down at the same time. The same is true of the direct proportion.

In a seesaw, as one end goes up, the other must go down, and vice versa. Likewise, in an inverse proportion, if one variable increases, the other must go down proportionally.

Why do some variables have this unique up-down proportional relationship? The answer has to do with *constant product,* the fact that when the paired quantities of the two variables are multiplied together, they always give the same product.

The constant product controls the relationship. It allows variation of the two variables only as long as the constant product is maintained. Mathematicians call this relationship *indirect variation* while the physical sciences use the term *inverse proportion.*

In the remainder of this chapter, some examples of inverse proportions will be examined, and the mathematics of working with inverse proportions will be studied.

4.2 SOME CASES OF INVERSE PROPORTIONS

A. A Vat of Juice

The first example of an inverse proportion to be examined involves a bottling company. Suppose that the bottling company is preparing to bottle a 240-litre vat of apple juice. The juice can either be packed into many containers with a little juice in each or else can be packed into a few containers with a large quantity of juice in each. The total quantity of juice has to be split up among all the bottles whether large or small.

Since the total quantity of juice is constant–it remains the same throughout–you can guess that an inverse (or direct) proportion exists between the number of containers and the amount of juice in each.

Using the constant product idea, complete the table below where V is the volume of a juice container in litres and C is the number of containers. In each case, C multiplied by V must equal the constant quantity of 240 litres.

V (litres)	10	4	2	1	$\frac{1}{2}$	$\frac{1}{4}$	$\frac{1}{10}$
C (containers)	24			240	480		2400

We see that as the volume of the containers becomes smaller, the number of bottles or other containers needed becomes bigger. The product of the two variables V and C is always the same; it equals the 240 litres that have to be divided up among the containers.

The general equation for this relationship is $VC = k$. It is the equation for an inverse proportion.

Of course, units must always be consistent. Are they consistent in $VC = k$?

V	C	=	k
$\frac{\text{volume}}{\text{container}}$	number of containers	=	total volume

Yes, the units are consistent.

(However, they may not lead to meaningful units in the answer as they do for direct proportions, and the method of units analysis is not very helpful for setting up solutions. This is largely because the units may not be chosen with the idea in mind of using analysis of units. Notice that we have chosen, in the above, to measure V in units of volume per container rather than just in units of volume, so that, by units analysis, k represents the constant volume. If, however, V were measured only in units of volume such as litres, then k becomes the product of total volume and number of containers is measured in litre-containers, a unit that is meaningless to us. The fact that one variable is actually a rate often escapes attention in inverse proportions.)

Now let us examine a second case of an inverse proportion.

B. Drugstore Pills

For our next example, let us consider the case of one hundred fifty grams of medication to be dispensed in pills. The medication can be divided into many pills each containing a little medication or into a few pills each with a large quantity of medication. The number of pills and mass of a pill are the variables in this relationship, and the total mass of the pills is constant no matter what the mass is of each pill. As the mass of medication per pill increases, the number of pills decreases to hold the product constant.

In the table below, P is the number of pills and M the mass of medication per pill. Fill in the missing quantities and write the equation for the relationship.

P (pills)	150	120				
M (g/pill)	1	1¼	1½	2	2½	5

The answers are given at the bottom of the page.

P (pills)	150	120	100	75	60	30
M (g/pill)	1	1¼	1½	2	2½	5

The constant product is 150 grams.

$PM = k$ where P = number of pills,
M = grams/pill,
k = total mass.

To fill in the first blank: $P = ?$, $M = 1\frac{1}{4}$ g/pill, $k = 150$ g. Then $(P)(1\frac{1}{4}\text{ g/pill}) = 150$ g:

$$P = \frac{150\text{ g}}{1\frac{1}{4}\text{ g/pill}} = 100\text{ pills.}$$

C. Worker-Days

Sometimes, the constant quantity that is being divided into parts is not as obvious as in the previous cases. Consider the case of 12 workers who take 180 days to complete a job. Suppose the number of workers is increased. What happens to the time required?

It should take more time with fewer workers. Conversely, it should take less time when there are more workers. Hence, this shows an inverse relationship. How can we tell if this is also an inverse (or indirect) proportion? If we can find a constant product for workers and days, then the variables are inversely proportional.

Although it may not be easy to recognize at first (the units don't help), it is the quantity of total work to be done—the job—that is being split up among the workers. Hence, it is the total work that is constant while the number of persons varies indirectly with the quantity of work per person. Complete the following where W = quantity of workers and D = time in days.

W	12	16	20		
D	180			90	40

Notice that the unit for W is given as workers and that the unit for D is days. Hence, the unit for the constant product is worker-days. A unit frequently in use for this kind of constant is man-hours. There is no convenient unit otherwise available for the "job." (The lack of familiar or meaningful units for the constant product is what prevents the convenient use of units analysis for analyzing many inverse (or indirect) proportions.)

It is certainly not immediately obvious that W and D are inversely proportional. In fact, you are not expected to "know" it just from the names of the variables. In science, you will be expected to recognize inversely proportional variables only when aided by tables of data or by equations. This will be discussed in later sections of this chapter.

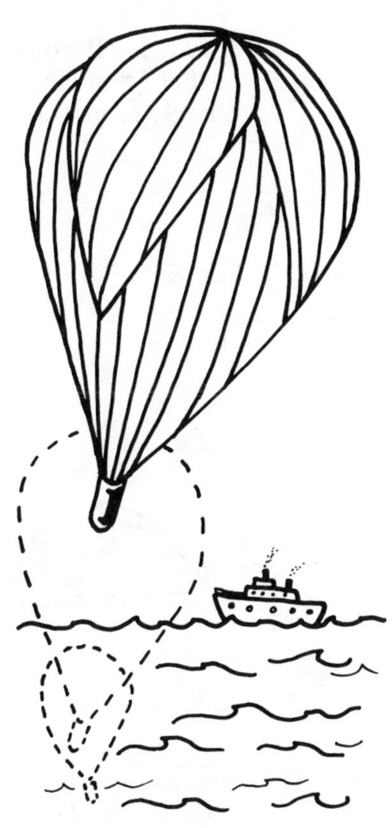

D. Pressure and Volume

The next case to be considered is that of pressure and volume. A balloon has a volume of 500 mL at a pressure of one atmosphere. It is taken underwater and gradually lowered into the ocean. As the pressure on it increases, it shrinks proportionately in volume. Show the mathematical relationship between pressure and volume.

Volume (mL)	500				
Pressure (atm.)	1	1.1	1.5	2.0	10

The above is an important example that you will encounter is science—two variables connected by a constant product.*

E. Inverse Money

The most common example of an inverse proportion is probably found in the spending of money. The *more* you spend per item, the fewer items you can proportionately buy for the same total amount of money.

What is the quantity of the constant product in the example?

Let I = number of items, and M = cost of an item. Write an equation to describe this inverse proportion. Use it to calculate M for 25 items.

*The meaning of the constant product is unknown even to some of the people who work with these variables in science and aeronautics. In fact, it takes an analysis of the units to show that the constant product of pressure and volume measures the energy of a system.

F. A Short-Cut

In the examples of inverse proportions discussed in Sections A to E, you first determined the constant product, k, which equals yx. Then, given y, you could solve for x or, given x, you could solve for y.

If one unknown is to be solved for, it is not even necessary to first determine the constant product. *Since the product of each pair of variables is always constant,* you can simply equate two inversely proportional pairs of variables. If $500 buys 4 items, to find out how many items can be purchased for $150:

$$(\$500)\ (4 \text{ items}) = (\$150)\ (x)$$

$$x = \frac{(500)\ (4 \text{ items})}{\$150} = 13.3 \text{ items.}$$

Use this method in the next problems, all of which involve inverse proportions.

1. Given $yx = k$, if $x = 6$ when $y = 14$, what is x when $y = 20$?

2. If $\frac{3}{4}$ gallon of ice cream is needed to fill each of 30 containers, how many containers of one gallon capacity each are needed to package the same total quantity of ice cream?

3. Given that 1000 Vitamin B pills are manufactured containing 0.062 g Vitamin B in a pill, how many pills can be produced if each pill instead contains 0.044 g Vitamin B?

4. If it takes 12 sewing machine operators to complete 60 dresses in one day, how many operators will it take to complete 60 dresses in 4 days?

5. Given a pressure of 2 atmospheres for a volume of 5 litres of gas, what is the volume of the gas at 3.8 atmospheres?

6. If I can afford to buy 30 ounces of peanut brittle at 45¢ per ounce, how many ounces of brittle can I buy for the same money at 30¢ per ounce?

4.3 IDENTIFICATION OF PROPORTIONS FROM TABLES OF DATA

For each of the following, state whether the variables are related by a direct or an indirect (inverse) proportion, or are not related in either way. Keep the following in mind:

Direct Proportion—Invariant Rate

Inverse Proportion—Constant Product

1.

Speed (mph)	Time (hours)
120	6
60	12
30	24
10	72

2.

Height (cm)	Width (cm)
8	2
4	1
24	6
20	5

3.

Pressure (lb./in.^{2})	Area (in.^{2})
4	20
8	10
10	8
16	5

4.

Number of Adults	Number of Children
10	30
20	20
25	15
5	35

5.

Salary ($/hr.)	Time (hrs.)
5	100
25	20
½	10
50	1000

6.

Volume (litres)	Height (cm)
30	120
3	1200
10	360
45	80
60	60

4.4 THE RECIPROCAL CHANGE

A. Variation of the Variables

We have so far examined the identification of an inverse proportion through the constant quantity that controls the relationship between the two variables. Next, let us look more closely at the way the two variables change.

We shall start with the seesaw equation for an inverse proportion, $yx = k$.

In this equation, as y gets bigger, x must get proportionately smaller in order to keep k at the same value. If y gets smaller, x will have to get proportionately bigger.

Suppose y becomes 5 times as big (increases to $5y$). What must happen to hold k constant? The equation below shows what happens, since the product of y and x always equals the constant, k.

$yx = k$

$5y \cdot (?)\, x = k$

$5y \cdot (\frac{1}{5})\, x = k.$

The equation shows that if y becomes 5 times as big, x becomes $\frac{1}{5}$ as big, changing to $\frac{1}{5}x$. Suppose y is initially 20 and x is 10, then

$20 \cdot 10 = 200.$

When y becomes 5 times as big, $5 \times 20 = 100$, so:

$100 \cdot ? = 200.$

Then 10 must reduce to 2, or become $\frac{1}{5}$ as big in order to maintain the constant product of 200.

$100 \cdot 2 = 200.$

Suppose that y instead becomes $\frac{1}{4}$ as big as it was before:

$yx = k$

$(\frac{1}{4}y) \cdot (?)\, x = k$

$(\frac{1}{4}y) \cdot (4)x = k.$

The above shows that if y becomes $\frac{1}{4}$ of what it was before, x must become 4 times as big. Suppose y is initially 20 and x is 10; if y becomes $\frac{1}{4}$ as big, which makes it equal to 5, then x becomes four times as big, which makes it equal to 40, and the product of the two is still 200 in both cases. That is:

$$yx = k$$
$$20 \times 10 = 200$$
$$5 \times 40 = 200.$$

Suppose y becomes q times as big:

$$qy \cdot (\,?\,)\,x = k$$
$$qy \cdot (\frac{1}{q})\,x = k.$$

This shows that whenever y *changes by the factor* q, x *changes by* $\frac{1}{q}$.

Likewise, if x changes by q, y changes by $\frac{1}{q}$.

Another way of saying this same rule is that:

> ***In an inverse proportion, as one variable changes by any factor, the other changes by the reciprocal of that factor.***

Do you recall the meaning of reciprocal?

The reciprocal of any number is "one" divided by that number. The reciprocal of five is $\frac{1}{5}$, of 3 is $\frac{1}{3}$, of 100 is $\frac{1}{100}$. To coin a word, the reciprocal of any number is "one-numberth" of what it was. The reciprocal of q is $\frac{1}{q}$.

B. Reciprocals

Before continuing the study of inverse proportions, here is some practice on calculating reciprocals. To take the reciprocal of any number, divide it into one. Thus, the reciprocal of 10 is $\frac{1}{10}$. To take the reciprocal of a fraction, recall that the procedure for a fraction in the denominator is to invert and multiply. The reciprocal of $\frac{1}{5}$ is

$$\frac{1}{\frac{1}{5}} = 1 \times \frac{5}{1} = 5.$$

To take the reciprocal of a decimal number, you can divide 1 by it. Another way is to convert the decimal to the corresponding fraction and invert.

For example, the reciprocal of 0.75, which equals $\frac{75}{100}$, is $\frac{100}{75}$ or $1\frac{1}{3}$.*

Likewise the reciprocal of 0.53 is $\frac{100}{53}$, of $\frac{1}{10}$ is 10, of $\frac{2}{50}$ is 25, and so on.

1. Find the reciprocal of each number:

   a. 25

   b. $\frac{1}{25}$

   c. $\frac{3}{10}$

   d. 5

2. Find the reciprocal of each number:

   a. 0.25

   b. $1\frac{1}{3}$

   c. 0.9

   d. 0.172

3. State which of the following are reciprocals of each other:

   a. 20 and $\frac{1}{20}$

   b. 20 and 0.05

   c. 0.25 and $\frac{1}{4}$

   d. $\frac{1}{5}$ and 5

---

*To show why this works, note that $\frac{1}{0.75} = \frac{1}{\frac{75}{100}}$. Multiply this by $\frac{\frac{100}{75}}{\frac{100}{75}}$, which equals 1.

$$\frac{1}{\frac{75}{100}} \times \frac{\frac{100}{75}}{\frac{100}{75}} = \frac{100}{75} = 1\frac{1}{3}.$$

e. $\dfrac{1}{\frac{1}{5}}$ and 5

f. 0.25 and 4

g. 100 and 10

h. $\dfrac{1}{200}$ and 0.02

4.5 IDENTIFYING THE RECIPROCAL FACTOR

The following exercise will help you see how the reciprocal factor of change operates. Notice, too, how it illustrates the seesaw of the inverse proportion.

Below is an example of related pairs of quantities for an inverse proportion. In this example, we have the case of a marathon runner practicing a measured route or track. The variables are number of laps required and length of track; the constant distance to be run is 24 miles.

Laps	Track Length (per lap)
96	¼ mile
48	½ mile
24	1 mile
12	2 miles
8	3 miles
6	4 miles

$$96 \text{ laps} \times \frac{\frac{1}{4}\text{ mile}}{\text{lap}} = 24 \text{ miles}$$

$$48 \text{ laps} \times \frac{\frac{1}{2}\text{ mile}}{\text{lap}} = 24 \text{ miles}$$

$$24 \text{ laps} \times \frac{1\text{ mile}}{\text{lap}} = 24 \text{ miles}$$

$$12 \text{ laps} \times \frac{2\text{ miles}}{\text{lap}} = 24 \text{ miles}$$

$$8 \text{ laps} \times \frac{3\text{ miles}}{\text{lap}} = 24 \text{ miles}$$

$$6 \text{ laps} \times \frac{4\text{ miles}}{\text{lap}} = 24 \text{ miles}$$

We see that as the track gets *shorter,* the number of laps needed *increases.* Therefore, the variables are inversely related. To show that the relationship is not only inverse but an inverse proportion, we note that the constant product is 24 miles.

The data should illustrate the reciprocal factor of change as well as the constant product. If one variable changes by any factor, the other variable should change by the reciprocal of that factor. For example, consider changing the laps from 96 to 12; the track length must be correspondingly increased from $\frac{1}{4}$ mile to 2 miles.

$$\frac{96 \text{ laps}}{12 \text{ laps}} = 8$$

$$\frac{\frac{1}{4}\text{ mile}}{2 \text{ miles}} = \frac{1}{8}.$$

Since 8 and $\frac{1}{8}$ are reciprocals, the two variables are inversely proportional to each other.

Show that the factor of change for laps is the reciprocal for that of track length in the problems below for the starting quantities of each variable and the changed quantities shown in (a) of the problem. Then repeat for the starting quantities of each variable and the changed quantities shown in (b).

1. 96 laps, ¼ mile and
   a. 24 laps, 1 mile
   b. 8 laps, 3 miles
2. 48 laps, ½ mile and
   a. 12 laps, 2 miles
   b. 96 laps, ¼ mile
3. 6 laps, 4 miles and
   a. 48 laps, ½ mile
   b. 24 laps, 1 mile

To illustrate the reciprocal relation in Problems 1 to 3 above, the factors of change are given in the following table.

Problem	Factor (laps)	Factor (distance)
1.a.	$\frac{1}{4}$	4
1.b.	$\frac{1}{12}$	12
2.a.	$\frac{1}{4}$	4
2.b.	2	$\frac{1}{2}$
3.a.	8	$\frac{1}{8}$
3.b.	4	$\frac{1}{4}$

Note that each factor of change is the reciprocal of the other factor, in either direction. If L is number of laps and D is distance then

$LD = k$

and if L changes by any factor, q, D changes by the reciprocal of that factor, $\frac{1}{q}$.

Here are two other examples of the reciprocal factor:

height = x height = $(1/3)x$

When three times as many people sit on a block of foam rubber, it lowers to $\frac{1}{3}$ the height.

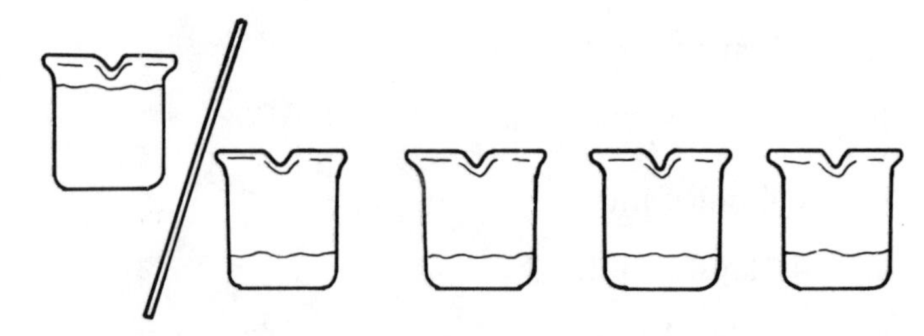

Water is poured into a beaker to a height of 8 cm. When the water from that beaker is poured in equal amounts into four other beakers, each holds 2 cm of water or ¼ of what it was before.

4.6 IDENTIFYING THE RELATIONSHIP

Now that you are more familiar with the nature of the inverse proportion and with some examples of it, you are ready to identify some relationships from the data.

When the variables are directly proportional,

1. the rate is invariant, and
2. they change by the same factor.

Either or both criteria may be used.

When the variables are inversely proportional,

1. the product is constant, and
2. they change by reciprocal factors.

Either or both criteria may be used.

Given data, you can identify whether variables are related, whether the relationship is direct or inverse, and whether a proportion exists.

State which of the following is the relationship for each pair of variables in the problems below.

a. Direct relation but not direct proportion.
b. Direct proportion.
c. Inverse relation but not inverse proportion.
d. Inverse proportion.

1. One dozen shampoo packages, 5 oz. weight; $5\frac{1}{2}$ dozen shampoo packages, 27.5 oz. weight.

2. A car driven at 20 miles per hour can brake to a stop in 12 feet. When the car is driven at 30 miles per hour, it needs 54 feet to brake to a stop.

3. It takes 9 people 22 days to do a job; 3 people need 66 days for the same job.

4. There are 3 g silver nitrate in 50 mL water and 18 g silver nitrate in 300 mL water.

5. At one gas station, I can buy 5 gallons gas at $1.80/gallon with the money I have. At the other station, I can buy $5\frac{1}{4}$ gallons for $9.45.

6. When apples weigh $\frac{1}{4}$ lb. each, there are 4 to a pound.

   When they weigh $\frac{1}{3}$ lb. each, there are 3 to a pound.

7. The electric train takes one minute to circle the track at 2000 feet per hour and two minutes at 1000 feet per hour.

4.7 SETTING UP EQUATIONS

Would you believe that you can now set up equations given the data for variables related by direct or inverse proportions?

Whenever the data show two variables to be directly proportional, you can write an equation of the form $y/x = a$ (the rate is invariant). If they are inversely proportional, then write that $yx = k$ (the product is constant). You can check direct or inverse proportionality by the constant factor or reciprocal factor of change.

The following show variables that are either directly or inversely proportional.

1. Plans are being made for a rectangular building with 12,000 square feet of floor space. The sides of the building can be lengthened or shortened in various combinations to give the floor space.

   For the same area, as length increases, width decreases.

Length (feet)	Width (feet)	Length × Width ($ft.^2$)
120	100	12,000
200	60	12,000
150	80	12,000

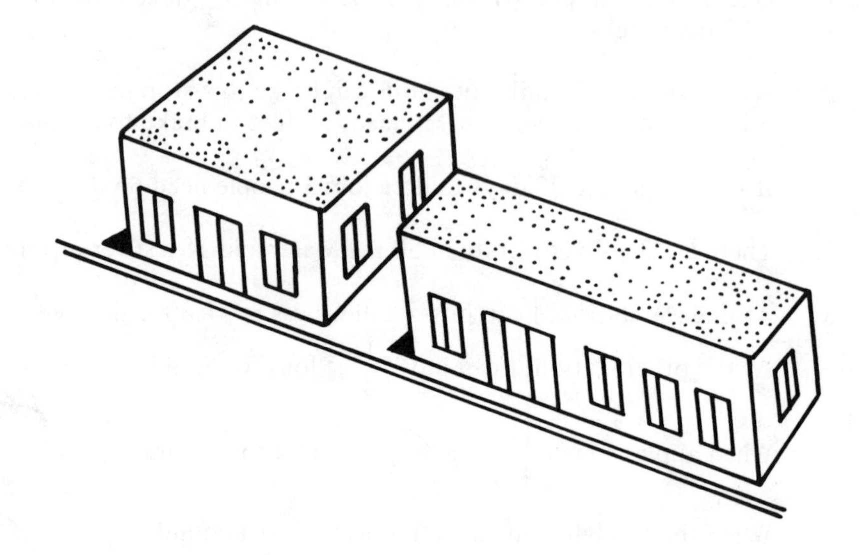

a. Set up the equation for the relationship.

b. Use your equation to find the length for a width of 400 feet.

c. What is the meaning of k, the constant product, and what is its unit of measurement?

2. The pressure inside the cylinder of an automobile engine is inversely proportional to its volume. Write the equation and use it for the following.
   If the pressure at the start of the upward stroke is 15 lbs./$in.^2$ for a volume of 38 $in.^3$, what is the pressure at the top of the stroke when the volume is $9\frac{1}{2}$ $in.^3$?

3.

Diameter (cm)	Length (cm)
20	10
18	11.1
10	20
2	100

a. Set up the equation for this relationship.

b. Use your equation to find the diameter when length = 23 cm.

4.

Time (hr.)	Distance (miles)
1	6
2	12
3	18

a. Set up the equation for this relationship.

b. Use your equation to solve for distance when time = 32 hours.

5.

Number of Pearls	Diameter of Pearls (mm)
20	5
15	6.67
10	10

a. Set up the equation for this relationship.

b. Use your equation to find how many pearls 8 mm in diameter can be strung on this length necklace.

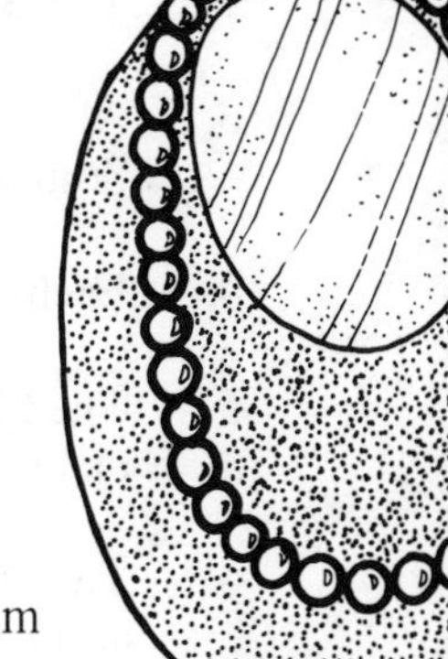

6.

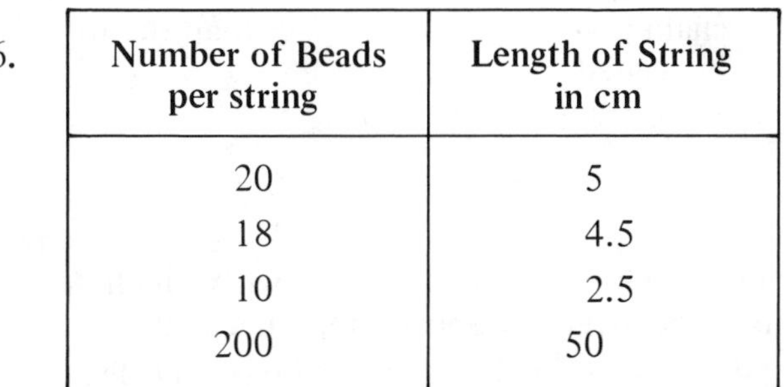

Number of Beads per string	Length of String in cm
20	5
18	4.5
10	2.5
200	50

a. Set up the equation for this relationship.

b. Use your equation to find the length of a string with 55 beads.

4.8 INVERSE BUT NOT PROPORTIONAL: THE CONSTANT SUM AND OTHERS

So much attention has been paid to proportional changes so far that one might think all changes are proportional. This is not so.

We will briefly examine one type of inverse relationship, the constant sum. Whenever two variables add up to a constant sum, the variables are inversely related but there is no proportional relationship.

Suppose Mr. and Mrs. Spunk together earn \$1000 per week. They spend some, S, and have the remainder, s. The total must be \$1000. The equation is:

$k = S + s$ where k = total earned = \$1000.

As S increases, s decreases and vice versa.

There are also many other types of relationships, some of which you may know from your earlier arithmetic courses.

In the following examples, first determine for each problem whether it is a case of direct or inverse proportion or is a constant sum or difference problem. Then, set up an equation for each problem showing the relationship between all the variables, and identify the symbols.

1. Plans are being made for a rectangular building with a perimeter (total length of the four sides) to equal 250 feet. Variations in the lengths of the sides are being considered.

2. There are 40 beads to a string. Some are red and the remainder are blue.

3. All of the pieces in the jigsaw puzzle are tossed onto a table and flattened out. Some are right-side up and some are wrong-side up.

4. Each gold chain has 25 links in it. The more chains, the more links.

5. The bigger the length of the link, the fewer links are needed for a 50-cm gold chain.

6. The chain is made up of 3 gold links, then a small diamond, then 3 gold links, and so on. Given the number of gold links, you can calculate the number of diamonds needed.

7. The length of the chain is always 0.6 cm longer than the total length of the links because of the clasp needed.

4.9 LAST WORDS

This chapter has expanded on the kinds of relationships with which you can deal. In addition to the direct proportion, the inverse proportion has been added to the kit of mental tools and understandings you have for "coping." In an inverse proportion between two variables, y and x, as x goes up, y goes down proportionately. The equation that expresses this is $yx = k$; it describes the seesaw relationship between y and x. An inverse proportion can be identified by the constant product of the paired variables and/or by the fact that when one variable changes by any factor or ratio, the other changes by the reciprocal of that factor.

There are other kinds of inverse relationships besides the inverse proportion. In each case, one variable gets bigger as the other gets smaller. One kind is the constant sum; x and y add up to the same quantity.

All relationships must be direct or inverse, or be a sequence of direct and inverse. Some, but not all relationships, are proportions. If there is any regularity in such relationships, an equation can be found to describe them and to enable further work to be done with them. Mathematics is a very powerful tool for us to use.

DEFINITIONS

Inverse proportion: An inverse proportion exists between two variables when the product of each pair of quantities for the two variables is constant. Also, if one variable changes by any factor, the other changes by the reciprocal of that factor.

Multiple: Same as factor (see "Definitions," Chapter One).

Reciprocal: The multiplicative inverse of a number or quantity, or the quotient of one divided by the number or quantity.

ACTIVITY 4.1–Measuring Sticks

Purpose: To find the relationship between lengths of sticks and the number of sticks needed to measure a fixed length.

Equipment: Sticks or rectangular blocks of three different lengths, a strip of masking tape stretched along a table top, and a cm ruler.

Procedure:

1. Using the shortest stick or block as a unit of length, measure the length of the strip on the table top. Repeat the measurement using the medium length stick or block. Then, repeat with the longest measuring stick. Record your results on the Report Sheet under (1).

2. Measure the length of each of the three sticks or blocks used, in centimeters; record under (2).

3. Answer the remaining questions on the Report Sheet.

ACTIVITY 4.2–What Is the Constant Relationship?

Purpose: To examine the relationship of length, width, and area of the rectangle.

Equipment: Graph paper, ruler.

Discussion: You wish to use 600 square tiles to construct a walk. The walk can be any constant width you use; of course, the length will depend upon the width you select. You will use graph paper to simulate the various possibilities so as to arrive at a decision.

Procedure:

1. Select 5 different widths of walk that you might desire, such as 5 tiles wide, 20 tiles wide, or any other number that divides evenly into 600. Use graph paper, and allow one tile to 1 or ½ small graph box as convenient. Mark the width and length on the paper of each of the 5 paths by counting off rows of the needed width for each until you have reached the equivalent of 600 tiles.

2. Count the number of tiles per row of length for each of the paths. Enter lengths and widths on the Report Sheet (2).

3. Complete the Report Sheet.

5

Meaningful Mathematical Equations

an advanced treatment

5.1 INTRODUCTION

Now that you have mastered the concepts of direct and inverse proportions and the way that they can be represented algebraically, you are ready to extend this mastery to reading all kinds of equations and to the understanding of how equations show relationships. Equations should be more than just something you memorize. They should be statements that have meaning for you, that you understand.

You have been given a great deal of drill on the mathematics used so far. This was necessary so that you would be ready to move onto these more abstract ideas.

Although the symbols to be used in this section are algebraic, they represent real-life variables, such as articles of merchandise, chemical molecules to be counted, lengths, volumes, time periods, pressures, voltages, or other variables that must be measured.

5.2 MATHEMATICAL SHORTHAND

Mathematical equations are shorthand statements of some often wordy paragraphs. For example, let's look at

$$S = \frac{D}{T}$$

S = speed
D = distance
T = time.

This equation shows us that when the rate of speed is held constant, distance and time are directly proportional to each other. For example, at 30 mi./hr., as time changes by any factor, the distance covered in that time likewise changes.

By rearranging the equation to

$$T = \frac{D}{S}$$

the equation can be considered from a different point of view. Suppose that instead of holding the speed constant and considering the relationship between distance covered and time required for an object moving at a given speed, we hold time constant. In that case, distance and speed become the variables. The equation shows that distance and speed are directly proportional. In a given time, if speed is doubled, so is distance. If speed changes by any factor, so does distance covered. Of course, this makes sense to us from our own experience.

In the following equations, state in words the relation between the variable on the left side of the equation and each of those on the right. Illustrate them with an example or statement invented by you to give them meaning.

Example: $W = st$

W = wages (earnings over total time)
s = salary (earnings per single time period)
t = time (total time).

When salary is constant, wages and time are directly proportional. This means that the longer you work, the more you earn. Also, when time is constant, wages and salary are directly proportional; the higher the salary, the more you earn proportionally for the same time worked.

1. $m = dv$

   d = density
   m = mass
   v = volume.

2. $C = PQ$

   C = total cost
   P = price/item
   Q = quantity of items.

3. $M = CF$

   M = length in meters
   F = length in feet
   C = conversion factor.

   (*Note:* You are restricted by reality as to how you can read this equation since C may have only one real value.)

4. $W = CV$

   W = weight of dissolved solid
   V = volume of solution
   C = concentration.

   Notice that you can derive the meaning of concentration from this equation if you don't already know it.

5.3 HOLDING ALL BUT TWO CONSTANT

Let us look more closely at the exercise that you just did. In this exercise, the relationships between the two variables were examined when the third was held constant. In fact, any time that you wish to know how a change in one variable affects other variables, you had best deal with only two variables at a time. If you keep all of the variables from changing (hold them constant) except for two of them, then you can see what happens to one variable as you change the other. The beauty of an equation is that it contains the complete information about the relationships between all the variables in it provided that you read it for two variables at a time.

The first rule for analyzing an equation follows.

> ***In order to see the connection between any two variables in an equation, hold all of the other variables constant. That is, allow no variables to undergo change except the two under examination.***

One more rule is all that is needed to analyze even complex equations. It is a rule you have already used for similar purposes. This second rule states that:

1. The quotient of two variables is invariant when they are directly proportional; $y/x = a$ or $y = ax$.
2. The product of two variables is constant when they are inversely (or indirectly) proportional; $yx = k$.

These two rules can be combined to analyze relationships in any equation. For example, consider the following formula for packaging ice cream for shipment to stores:

$$V = CN$$

where V = total volume of ice cream
C = volume per container
N = number of ice cream containers.

What is the relationship for each of the three possible pairs of variables: V and C; V and N; and C and N?

These are the total number of pairs of variables possible in the given equation.

We will consider each in turn.

V and C

Applying the first rule, hold N constant. We will start with some fixed (or unvarying or constant) quantity of containers to be shipped, and find out how the volume needed in each container varies (increases or decreases) as the total amount of ice cream varies. We can write $N = k$ where k is constant so that

$$V = kC.$$

Next, the second rule is applied. From the arrangement of the variables of this equation, we immediately know that V and C are directly proportional with all the meanings thereof (invariant rate where $V/C = k$; constant ratio of change where $qV = k(qC)$).

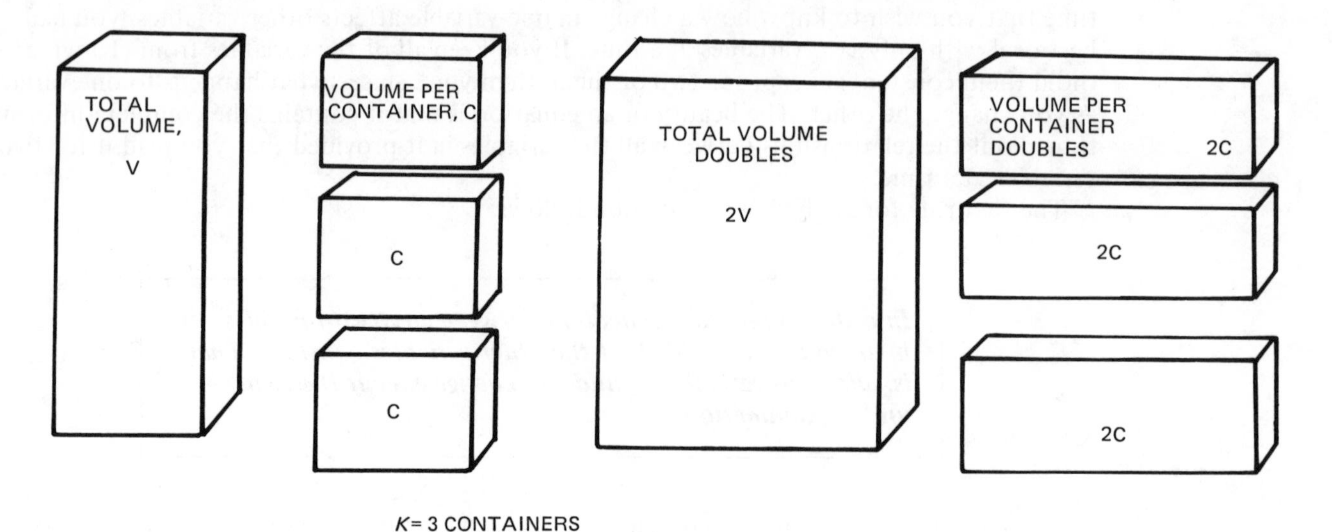

V and N

By holding C constant this time, we see that $V = kN$, or, in words, that the total volume of ice cream is directly proportional to the number of containers of a constant volume used.

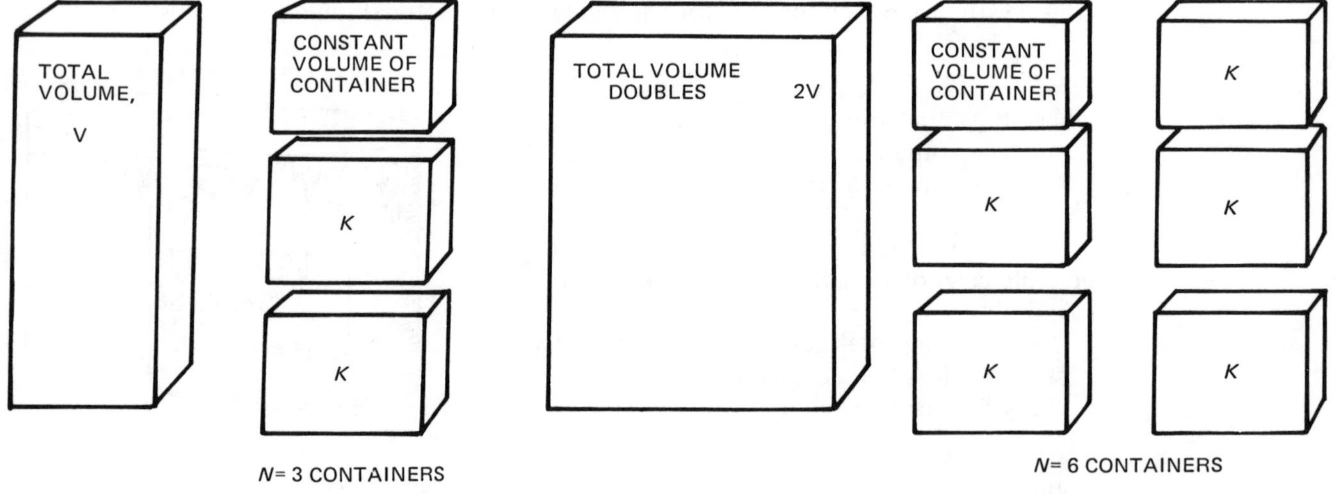

C and N

Holding V constant, the equation becomes $CN = k$. According to the second rule, C and N are inversely proportional. As the volume of a container increases, the number of containers needed to pack a given quantity of ice cream decreases. Conversely, as the container becomes smaller, the number of containers needed to pack up all the ice cream increases.

$$CN = (QC)\left(\frac{1}{Q}N\right) \quad \text{or} \quad CN = \left(\frac{1}{Q}C\right)(QN).$$

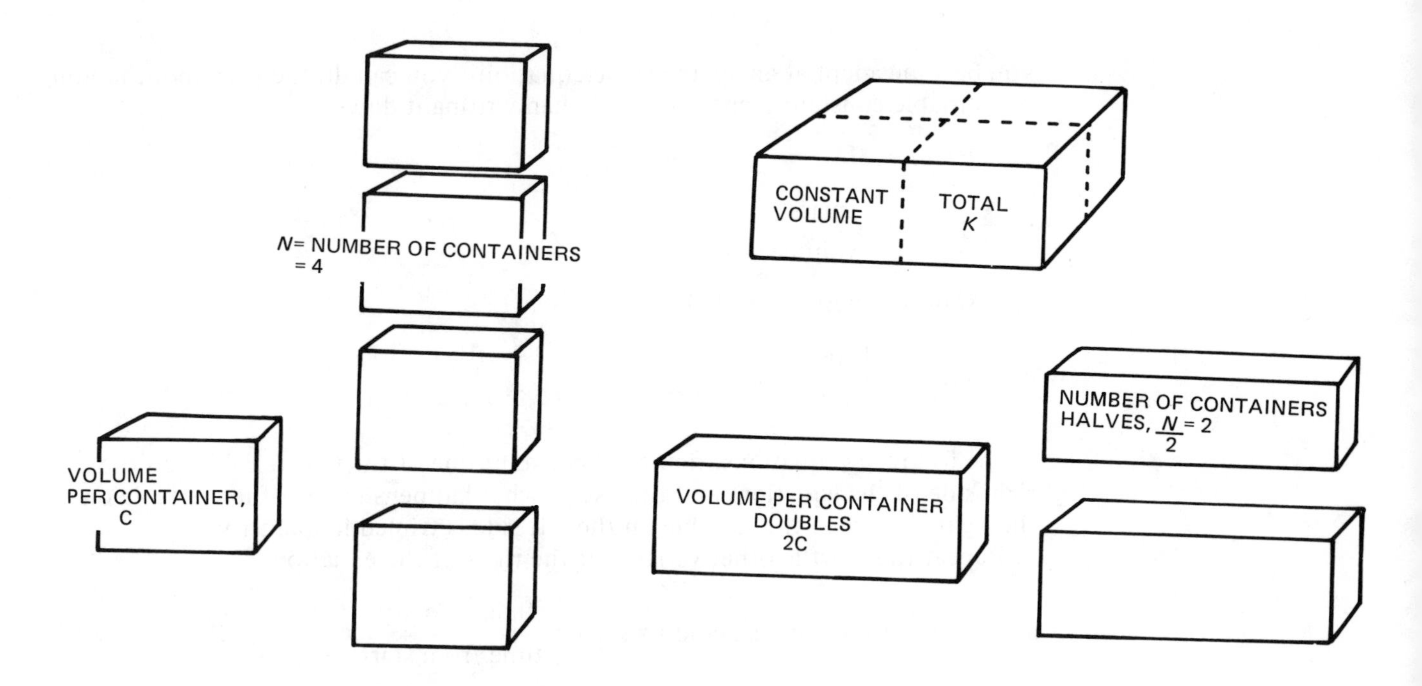

To summarize, for $V = CN$:

Held Constant	Proportional Change
C	$V\uparrow, N\uparrow; V\downarrow, N\downarrow$
N	$V\uparrow, C\uparrow; V\downarrow, C\downarrow$
V	$C\uparrow, N\downarrow; C\downarrow, N\uparrow$

The information in the above can be summarized in words by saying that V and N are directly proportional when C is held constant, that V and C are directly proportional when N is held constant, and that C and N are inversely proportional when V is held constant.

Or, you can summarize it all in the brief mathematical statement:

$V = CN.$

5.4 MEANINGS OF EQUATIONS THAT CONTAIN NO SUMS OR EXPONENTS

You now have sufficient information to analyze equations that have been established by gathering data on real relationships in science. Some examples of the types of equations taught in the introductory sciences are:

$v = at,\ \ pv = nRT,\ \ d = m/v.$

A. Equations with Three Variables

1. In each of the following, state which variables are directly proportional to each other and which are inversely proportional for all three possible combinations. As

you become adept at analyzing these equations, you can do the part about holding one variable constant mentally rather than writing it down.

a. $v = at$.

b. $\text{density} = \dfrac{\text{mass}}{\text{volume}}$.

c. squops = jiggers × swindlers.

d. $\text{speed} = \dfrac{\text{distance}}{\text{time}}$.

2. In the following equations, first hold constant one of the two variables on the right side of the equation, and then state what happens to the other variable on the right side when the variable on the left side is tripled in quantity.
   Repeat this for the other variable at the right of the equation.

a. $\text{acceleration from a standing start} = \dfrac{\text{final velocity}}{\text{time from start}}$.

b. distance = (speed) (time).

c. $d = m/v$.

d. $\text{percentage concentration} = \dfrac{\text{mass of solid}}{\text{mass of solution}} \times 100\%$.

B. Equations with Four or More Variables

When an equation has more than two variables, the relationships between any two variables can be analyzed by holding all of the other variables constant.
Some sample problems follow.

Sample Problem One

Problem: What is the relationship between x and y in $y = \dfrac{axc}{d}$?

Solution: Hold a, c and d all constant; allow $\dfrac{ac}{d}$ to equal k. Substituting, $y = kx$. The equation shows y and x are directly proportional.

Sample Problem Two

Problem: In the equation of Sample Problem One, what is the relationship between y and c?

Solution: Let $\dfrac{ax}{d} = k$. Then $y = kc$. Therefore, y and c are directly proportional.

Sample Problem Three

Problem: In the equation of Sample Problem One, what is the relationship between y and d?

Solution: Let $axc = k$. Then $y = \frac{k}{d}$. Therefore, y and d are inversely proportional.

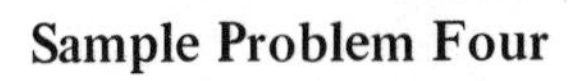

Sample Problem Four

Problem: In the equation of Sample Problem One, what is the relationship between a and d?

Solution: Rearrange the equation to get all of the quantities to be held constant on the same side:

$$\frac{a}{d} = \frac{y}{xc}.$$

(This step can be done mentally.)

Let $y/xc = k$.

Then $a/d = k$.

Therefore, a is directly proportional to d.

State the relationship between y and x in the following.

3. $yx = qr$.
4. $ayx = qrs$.
5. $ax = qry$.

In the following, assume all symbols stand for variables.

6. Given $y^2a = bc$, what is the relationship between a and b?
7. Given $\frac{y^2a}{f} = b^2cd$, what is the relationship between:
   a. a and d.
   b. a and f.
   c. f and d.
8. Given $3y^2 = bc$, what is the relationship between b and c?
9. Given $a^2bc = d^2e\sqrt{f}$, what is the relationship between:
   a. b and c.
   b. b and e.

10. Given $pV = nRT$ where p = pressure, V = volume, n = number of moles, R is a constant, and T = temperature, state the relationship in words between:

    a. p and V.

    b. p and T.

    c. V and T.

    d. n and T.

    e. V and n.

5.5 EQUATIONS WITH EXPONENTS

A. Complex Relationships

So far, we have studied the way two variables that are directly proportional can be related by an equation. Such equations, however, describe only some of the mathematical relationships useful in business, science, and other areas of application. How can we analyze more complex equations such as $y = kx^2$, $y = kx^3$, $y = k/x^2$, or $y = k\sqrt{x}$ so as to find meaning in them?

The remainder of this chapter will show some of the ways to make sensible to you certain equations that are not as simple as $y = ax$.

The method used will take advantage of the following idea. Starting with the equation $y = kx$, where y and x are proportional variables and k is a constant, suppose some other expression is substituted for x. Whatever that other expression is, it is proportional to y. For example, suppose the equation is $y = kab$ where k is constant. Then it is ab that is proportional to y. Or, if $y = kc/x$ where k is a constant, then c/x is proportional to y.

In the following, state what is proportional to y given that k is constant.

1. $y = kx^2$.

2. $y = kx^3$.

3. $y = k\sqrt{ac}$.

4. $xy = k$.

5. $xy = kb$.

B. Factors of Change

We saw earlier that when two variables are directly proportional to each other, whenever one changes by a certain factor or ratio, the other changes by the same factor. Suppose, however, that it is not x and y that are directly proportional but x and y^2. How do changes in x affect y and the converse?

The same ideas about factor of change that apply to equations of the form $y = kx$ may be applied to equations with exponents in them.

Consider the equation:

$y = kx^2$.

Inspection shows that y and x are not proportional. However, y and x^2 are directly proportional.

Answer the following for the equation $y = kx^2$.

6. If x^2 doubles to become $2x^2$, what happens to y?

7. If x^2 becomes ax^2, what happens to y?

8. Given $y = kx^2$, fill in the missing part:

   a. $(\quad)y = k(2x)^2$.

   b. $(\quad)y = k(3x)^2$.

   c. $(\quad)y = k(4x)^2$.

   d. $(\quad)y = k(5x)^2$.

   e. $(\quad)y = k(6)^2(x)^2$.

   f. $(\quad)y = k(ax)^2$.

   g. When $y = kx^2$ and x changes by the factor of a, by what factor does y change?

9. Fill in the table, given that $y = 2x^2$.

x	x^2	y
1		
2		
3		
4		
8		

   a. When x doubles from 1 to 2, by what factor does y change?

   b. When x doubles from 2 to 4, by what factor does y change?

   c. When x quadruples from 1 to 4, by what factor does y change?

   d. When x changes by a factor of 8 from 1 to 8, by what factor does y change?

   e. When x changes by a factor of a, by what factor does y change?

From the previous table, the rule can be stated for changes in y and x^2 in $y = kx^2$.

> ***Given*** $y = kx^2$, ***whenever*** x ***changes by the factor*** a, y ***changes by the factor*** a^2.

10. Given the equation $d = \frac{1}{2}at^2$, state how d changes when t:

    a. Doubles.

    b. Halves.

    c. Changes by a factor of 5.

    d. Changes by a factor of $\frac{1}{3}$.

11. Given the equation $d = \frac{1}{2}at^2$, when t changes from 5 to 15, how does d change?

12. Given the equation $y = kx^2$:

    a. What happens to y when x changes from 20 to 10?

    b. What happens to y when x changes from 10 to 30?

13. Given $y = \frac{k}{x^2}$ (or $yx^2 = k$), fill in the missing parts in the following:

    a. $(\quad)(y) = \frac{k}{(2x)^2}$.

    b. $(\quad)(y) = \frac{k}{(3x)^2}$.

    c. $(\quad)(y) = \frac{k}{(10x)^2}$.

    d. $(\quad)(y) = \frac{k}{(ax)^2}$.

14. Based on your answers to Problem 13, state the rule for the way y changes whenever x changes by the factor of a in the equation

    $y = \frac{k}{x^2}$.

15. Given the equation $yx^2 = k$, state how y must change when x changes by a factor of:

    a. 2.

    b. 5.

    c. $\frac{1}{2}$.

d. $\frac{1}{5}$.

e. 0.1.

f. 100.

16. Find the rule to show what happens to y when x changes by any factor in the equation $y = k\sqrt{x}$.

17. Given the equation $F = \frac{GmM}{d^2}$, state the relationship between:

a. F and M.

b. m and M.

c. M and d^2.

d. M and d.

e. F and d^2.

f. F and d.

5.6 ADDITIVE-SUBTRACTIVE EQUATIONS

A. The Equation, Simplest Form

Let us consider next an equation that has more than one term in it, such as $y = mx + b$. In an equation of this type, y gets bigger as x gets bigger and gets smaller as x gets smaller, so y and x are directly related but they are not proportional. As x changes by any factor, y does not change by the same factor.

In the simplest form of this equation, $m = 1$ in which case $y = x + b$. Again, let us consider only two variables at a time.

If b is held constant, then y is always b units bigger than x. For example, if John is always 2 years older than Jane, then b is 2 years (y = John's age and x = Jane's age). If a certain length is always 3 feet more than the width, then the length is y, the width is x, and b equals 3 feet. All the units must be the same for the equation to be an equality.

If y is held constant, the sum of x and b must always equal y. Then, y is a total made up of x and b. If a room holds 40 people, it may hold 10 women and 30 men, or 20 women and 20 men, or 35 women and 5 men. In each case, y = 40 people, x = number of women, and b = number of men (the units are people).

In the following problems, set up the general equation, identifying the symbols you select and giving the quantity of any constant.

Sample Problem

Problem: The average boy at a certain age is $1\frac{1}{2}$ inches taller than the average girl.

Solution: The equation is: $B = G + b$

where B = height of boy
G = height of girl
b = constant = $1\frac{1}{2}$ inches.

---

1. Alice earns \$50 more than does Belle each week.

2. The Captain Marvel High School team played 30 baseball games last year. They won 16 and lost 14.

3. There are 16 slices to a loaf of bread. When E are eaten, L are left.

4. The train is 2000 feet long. When the front of the train is 50 feet from the station, the back is 2050 feet away.

5. The two long sides plus the two widths of the playground must total 100 yards.

6. John rode the horse part of the way and walked it the rest of the way from the farm to the town.

B. The Equation: $y = mx + b$

The equation $y = mx + b$ can be rearranged to $y - b = mx$. Looking at it this way, it can be seen that if m is a constant and x changes by any factor, $(y - b)$ changes by the same factor. Thus, if b is subtracted from y, the remainder is proportional to x. One of the more familiar examples of this relationship is the conversion equation for °F and °C:

$°F - 32° = \frac{9}{5}°C$ (or $°F = \frac{9}{5}°C + 32°$).

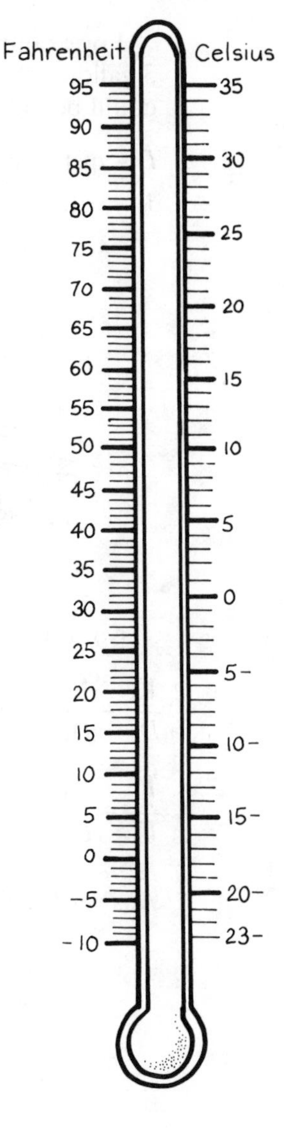

This equation says that if 32° are subtracted from the Fahrenheit reading, then each Fahrenheit remainder equals $\frac{9}{5}$ or 1.8 times as much as the Celsius reading.

7. A drawing of a thermometer graduated both in the Fahrenheit scale and the Celsius scale is shown. For each of the following, check that when you subtract 32° from the Fahrenheit scale and take $\frac{5}{9}$ of the remainder, it equals the corresponding reading on the Celsius side of the scale.

   a. 86°F.

   b. 40°F.

   c. 10°F.

8. Show that if the Celsius reading changes from 20° to 40°C, the Fahrenheit reading less 32° also doubles.

9. Show that if the Celsius reading changes from 10° to 30°, the Fahrenheit reading less 32° also triples.

10. Suggest a reason why 32° must be subtracted from the Fahrenheit reading in the conversion equation.

5.7 ADDITIONAL SHORTHAND

There is one more type of mathematical shorthand that you may encounter in applied arithmetic or in science that we haven't mentioned so far. This is the practice of labeling starting and final symbols for the quantities of a variable by subscripts, such as X_1, and X_2.

For example, if a motorcycle gets 175 miles to a gallon of gas and therefore 425 miles to 3 gallons, the variables are distance covered and volume of gasoline, and the quantities could be indicated by:

D = distance	$D_1 = 175$ miles	$V_1 = 1$ gallon
V = volume of gasoline	$D_2 = 425$ miles	$V_2 = 3$ gallons

Not only can the equation $V/D = k$ be written, but one can also write:

$\frac{V_2}{D_2} = \frac{V_1}{D_1}$ rate proportion, and

$\frac{V_2}{V_1} = \frac{D_2}{D_1}$ ratio proportion.

Answer the following:

1. In the next relationships shown, identify the two quantities for each variable by labeling them according to the system just described.

   a. $\frac{55 \text{ miles}}{\text{hour}} = \frac{110 \text{ miles}}{2 \text{ hours}}$.

   b. $\frac{2 \text{ quarts oil}}{10 \text{ quarts oil}} = \frac{\text{one car}}{\text{five cars}}$.

   c. $\frac{260 \text{ balls pitched}}{65 \text{ balls hit}} = \frac{100 \text{ balls pitched}}{25 \text{ balls hit}}$.

2. Use the same labeling system for the inverse proportion $yx = k$ and derive the equations corresponding to the ones obtained for a direct proportion.

3. Is $\frac{y_1}{y_2} = \frac{x_2}{x_1}$ an expression for a direct or inverse proportion?

4. Is $y_1 x_1 = y_2 x_2$ an expression for a direct or inverse proportion?

5. Write the symbols for starting and final quantities and the equations for Problems 4 through 6, Section 4.7.

6. Write the symbols for starting and final quantities and the equations for Problems 3 through 7 in Section 4.6.

5.8 LAST WORDS

This chapter, in its brief treatment, is intended to prepare the student for understanding relationships in equations that are encountered in certain parts of introductory science courses. By then, the student will have had additional preparation and many more applications in the laboratory and in written examples. Such equations are expected by then to be as comfortable to the student as those simple ones that describe unit pricing in the supermarket.

In this chapter, equations were analyzed to show which variables or expressions are proportional to each other.

ACTIVITY 5.1–The Law of the Lever

Purpose: To investigate the relationship between the variables of a simple lever.

Materials: Thread, 12-inch wooden ruler, large paper clips, thread support, masking tape.

Procedure: Tie a long piece of thread securely around the exact middle of the ruler so that the ruler balances horizontally when it hangs by the thread. If the ruler tilts when hung, place masking tape at the far end of the uptilted ruler to weigh down that side until it is horizontal. Attach the other end of the string to a support or hold it in your hand so that the ruler hangs horizontally and freely.

Slide one large paper clip onto the ruler until it is exactly over the 10-inch mark. Now, slide another paper clip over the opposite end of the ruler and move it along until the ruler is again exactly horizontal. Answer (1).

Next, slide a second paper clip next to the one on the ten inch mark, and then move them as a pair together along the ruler until it is again balanced. Answer (2), (3), and (4).

After you have answered (4), you are to determine whether the relationship is directly or inversely proportional. Briefly write on a piece of paper what activity you will do to find this out. After your teacher has approved of your plan, proceed to obtain the needed data and record what you have done on the Report Sheet under (5). Then, show your calculations in (6) and answer (7).

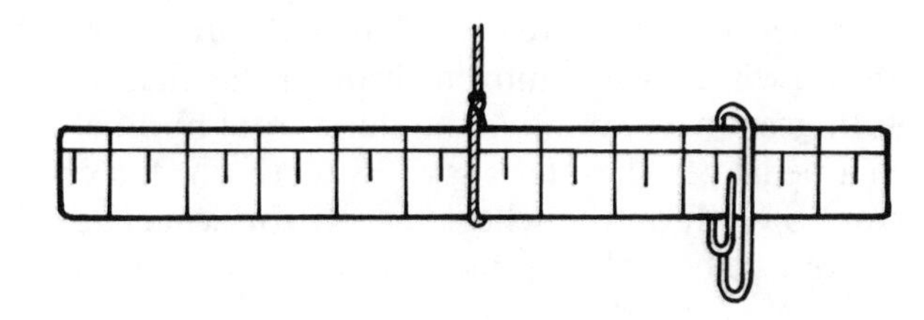

ACTIVITY 5.2–Bicycle Gear System

Purpose: To investigate how a gear system works on a 10-speed bicycle.

Equipment: Ten-speed bicycle with uncovered gear system.

Procedure:

1. Examine the system of gears on the bicycle. Turn it upside down to see the action more easily. You will find five toothed or notched discs on the rear wheel and two toothed discs on the pedal. The toothed discs are called gears. Check the following statements for yourself by observing the cycling of the bicycle. Every time the pedal makes one complete turn, each of the two gears attached to the pedal shaft also makes one complete turn.

   A sprocket chain fits on a loop over one rear wheel gear and one pedal gear. Every time the pedal is turned, the teeth on the gear move into the chain and cause the chain to move around. The back end of the chain pushes the teeth on the back wheel gear and causes the wheel to move. This is called gear action.

   Whenever the "speed" setting of the bicycle is changed, the chain is shifted to interlock with another gear.

   Work the gear action of the bicycle until you are familiar with the operation.

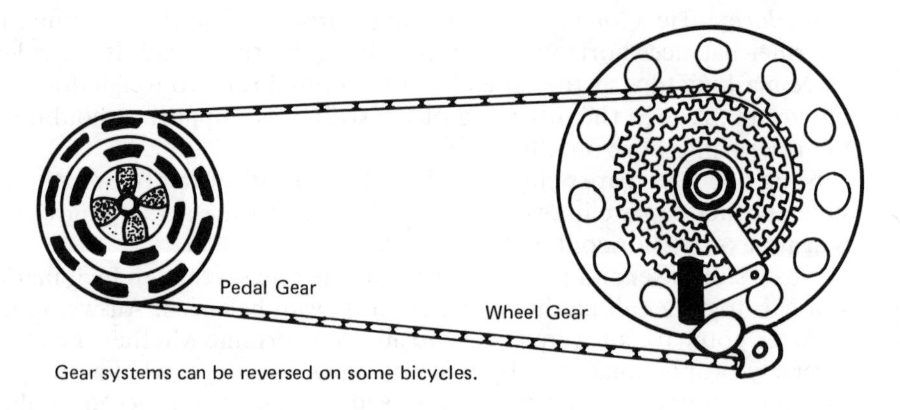

Gear systems can be reversed on some bicycles.

2. Count the teeth on each gear. Enter on the Report Sheet.

3. Since the chain is the same throughout the gear action, the same size links are pushed around and around. Every time one pedal gear tooth pushes through the chain, one wheel gear tooth must also be pushed through. Hence, when the foot pedal is turned so that its teeth are moved through the chain, the same number of teeth on the wheel gear is moved through the chain and so the wheel is forced to turn. Both gears have to move the same total number of teeth through the chain.

   When a pedal gear has, let's say, 28 teeth on it, a complete turn of the pedal will move 28 teeth on the wheel gear. If the wheel gear has only 14 teeth on it, it will make two complete rotations for the one rotation of the foot pedal.

Pedal Gear	Wheel Gear
28 teeth	14 teeth
1 turn	2 turns

We see that: teeth $\times$ turns = teeth $\times$ turns

and that teeth moved through the chain and the number of turns of the gear are inversely proportional for two gears connected through a chain. The specific equation for this system is

teeth on pedal gear $\times$ turns of pedal = teeth on wheel gear $\times$ turns of wheel.

Thus, if the teeth on pedal gear and on wheel gear are known, you can calculate the turns of the wheel for a given number of turns of the pedal. Answer (3).

4. Answer (4).

5. The number of turns of the wheel gear per one turn of the pedal gear is called the gear ratio:

$$\text{gear ratio} = \frac{\text{turns of wheel gear}}{\text{turns of pedal gear}}.$$

The answers that you have calculated for (3) are also the gear ratios. Answer (5).

6. Complete the remaining questions on the Report Sheet.

6

Analysis of Equations for Problem Solving

putting it all together

This chapter brings together all that has been developed so far in this module.

The following problems involve construction of equations for direct proportions, inverse proportions, and additive-subtractive relationships, and for analysis of all the relationships in several kinds of algebraic equations for real-world variables.

The strategies involved in such problems always require the same decisions:

1. recognition of variables and constants,
2. recognition of relationships, and
3. statement of relationships.

The order for 1. and 2. may sometimes reverse.

Answer the following.

1. Given the equation $C = PR$, where C is the total cost, R is the number of roses, and P is the price per rose, calculate P given that C equals \$20.70 and R equals 6 roses.

2. Only one of the tables below shows a directly proportional rate. State which is the table and prove your answer.

A	B
400	2
40	20
4	200

C	D
2.0	8
3.5	14
7.5	30

E	F
25	5
50	10
75	20

3. State whether each of the following is a case of:

   A. a direct proportion,

   B. a direct relation but not a direct proportion,

   C. an inverse proportion,

   D. an inverse relation but not an inverse proportion.

   a. An object falls 16 feet in one second and 64 feet in two seconds.

   b. A car travels 30 miles in $\frac{1}{2}$ hour and 300 miles in 5 hours.

   c. The volume of a bicycle tire is 2800 cm^3 at a pressure of 16 lbs./in.^{2}. At a pressure of 18 lbs./in.^{2}, it is 2489 cm^3.

   d. John is 8 years old when Michael is 10.

   e. It takes 3 people 18 days to construct a shed. Six people need 9 days for it.

   f. At 56 miles per hour, a car takes 4 hours to travel from Stamford to Boston. At 40 miles/hr., it takes 5.6 hours.

4. State whether the following variables are related by a direct proportion:

   a. y and x in the equation $y/x = 3$.

   b. y and s in the equation $y = st^2$.

   c. y and t in the equation $y = st^2$.

   d. s and t in the equation $ay = st$.

5. Write an equation in symbols and define each symbol for each invariant rate below:

   a. 84 lbs./38 kg.

   b. 11 players/team.

   c. 1 yard/$13.

   d. 15 oz. oil/4 oz. vinegar.

6. Given the equation, $G = kR$, where G = number of grooves and R = number of records, what are the units for k?

7. Write an equation to show the total weight of luggage to be expected on a passenger plane going from New York City to Chicago if the usual is 100 lbs. luggage for every 6 passengers.

8. Given the table of values below, state the general equation relating the two variables.

Steel (lbs.)	Cement (tons)
425	6.8
850	13.6
1275	20.4

9. Given the equation $q = rs$, where all variables are held constant except the two being considered, answer the following:

   a. What happens to q if r is quadrupled?

   b. What happens to s if q changes from 16 to 4?

   c. What happens to r if q changes from 5 to 25?

   d. What happens to r if q becomes $\frac{1}{10}$ as big?

   e. What happens to r if s becomes 10 times bigger?

   f. What happens to s if r becomes $\frac{1}{4}$ as big?

   g. If r is initially 10 and q changes from one to ten, what is the new value of r?

10. Complete the following tables for two variables related by an inverse proportion.

a.

x	y
5	50
2	
10	
$\frac{1}{5}$	

b.

q	r
400	10
80	
	40
4000	

c.

f	g
7	21
	28
$3\frac{1}{2}$	
28	

11. Given the equation $F = ma$,

    where F = the force acting on an object,
    M = mass of the object,
    a = acceleration of the object due to the force.

    Explain your answers to the following in terms of the above equation:

    a. As the force exerted on an object increases, does acceleration increase or decrease?

    b. To get an object to accelerate twice as fast, how should the force be changed?

    c. If the same force is exerted on two different cars, one a small car weighing 2000 lbs. and the other a light truck weighing 6000 lbs., compare the resulting acceleration for each.

12. Hartford is 42 miles from New Haven or 67.6 kilometers. Write a general equation in symbols to relate the two units of measurement to each other. Define the symbols. Use the equation to calculate the distance in miles for 23 kilometers.

13. Set up an equation relating the variables in each of the charts below.

a.

Distance (miles)	**Speed (mi./hr.)**
100	50
200	100
350	175

b.

Speed (mi./hr.)	**Time (hours)**
50	2
100	1
350	0.286

c.

Number of Blocks in a Box in a Row	**Length of Box (inches)**
100	2
150	$1\frac{1}{3}$
300	$\frac{2}{3}$

d.

Length A (km)	Length B (km)
40	45
31	36
55	60
2	7

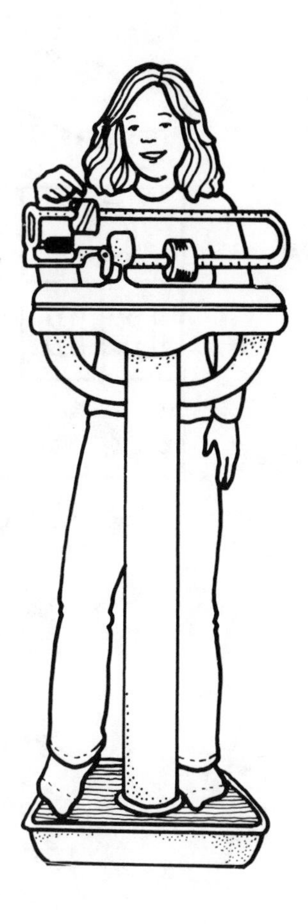

14. Alice weighs 100 pounds. Another scale shows her as 45.4 kg. Write a general equation in symbols that describes the relationship between the units; identify the symbols.

15. It takes 10 workers 40 days to build an ark. How long does it take 100 workers?

16. At 55 miles per hour, a car can travel from Milford to Florida in 28 hours. How long will it take at 50 miles per hour?

17. John earns \$750 in 5 weeks. Then his salary is tripled. How long does it take him to earn \$750 at the new rate?

18. When an object is immersed in a liquid, it may bob to the surface or may float in the liquid. The equation that describes this is $F = pgv$

    where v = volume of the object,
    p = density of fluid,
    g = the constant acceleration due to gravity,
    F = the force pushing the object up.

    a. When F is in dynes, p is in g/cm^3, g is in cm/sec^2, and v is in cm^3. What is one dyne equal to in metric units?

    b. As the volume of the object increases, is it more or less likely to float upwards? Explain, based on the equation.

    c. Is an object more likely to float or sink as the liquid gets denser? Mercury is denser than water; will an object be more likely to float in mercury or in water? Explain, based on the equation.

19. Given a solution that contains 25 g of solute per litre, the solution is boiled down, evaporating the solvent, until it contains 60 g of solute per litre. If you start with 0.5 litres of solution, how much solution is left at the end?

20. A set of play blocks is arranged as shown below. The length of each block is considered to be the distance measured from left to right as shown on this page.

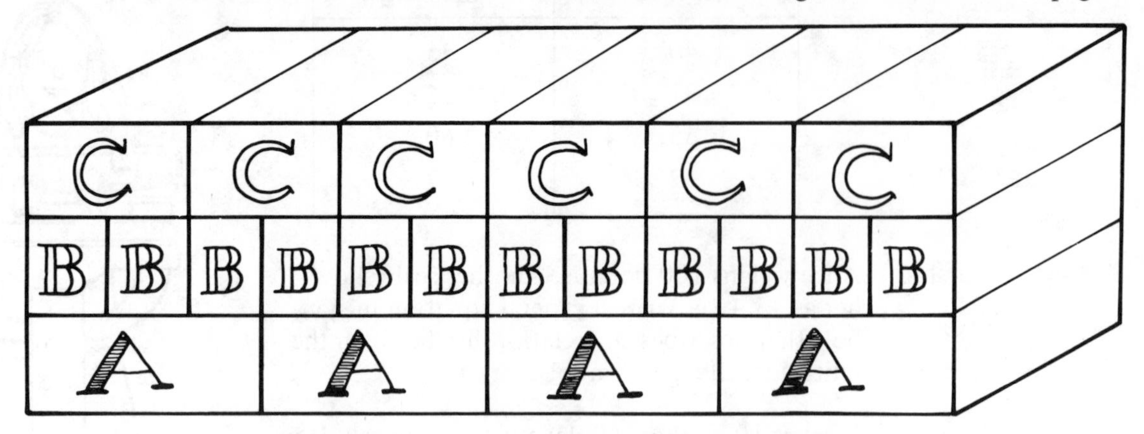

a. If L is the length of a block, N equal to the number of blocks, and W equal to the length of a row of identical blocks, write an equation to show the relationship between L, N, and W.

b. From your equation, determine how W changes if L increases while N is held constant.

c. Use your equation to find out how L changes when N decreases and W is held constant.

d. Convert your equation to use W_1, L_1 and W_2, L_2 for the variables when N is held constant.

e. Convert your equation to use N_1, L_1 and N_2, L_2 when W is held constant.

21. Write an equation to show each of the following relationships:

a. s is inversely proportional to t.

b. s is directly proportional to t^2.

22. Given the equation, $v_f = v_0 + at$, as t changes by any factor, what happens to v_f?

23. A racing car is gunned up to maximum speed with a constant acceleration all the way. The equation describing this is $d = \frac{1}{2}at^2$ where d is the total distance covered, a is the constant acceleration and t is the elapsed time (total time). As the time triples, what happens to the distance covered?

24. Pure gold is 24 karat gold. Given 100 g of pure gold, how much 18 karat gold can be made from it? (This is an inverse proportion if we assume that 18 karat gold is alloyed with a metal of the same density.)

25. The area of a circle is given by the equation $A = kr^2$, where A equals the area and r is the radius of the circle.

a. When the radius doubles in length, what happens to the area of the circle?

b. When the radius becomes four times as big as it was, what happens to the area of the circle?

c. In order for the area to become 25 times as big, what must happen to the radius?

d. When the radius becomes $\frac{1}{4}$ of what it was, what happens to the area?

26. Given the equation $F = \frac{kMm}{d^2}$, where F is the gravitational force, M and m are two masses being attracted toward each other by gravity, and d is the distance between the two masses, answer the following:

a. When the two masses get closer, so that the distance between them is only one-half of what it was, what happens to the gravitational force between them?

b. Suppose the two masses are instead pulled apart until they are three times as far as they were initially, what happens to the gravitational force between them?

7

Graphic Pictures

7.1 INTRODUCTION

This chapter will look into another way of showing relationships between variables, a way that is essentially pictorial. Graphs allow the nature of a relationship to be seen at a glance. They present a picture that offers much information to the trained eye.

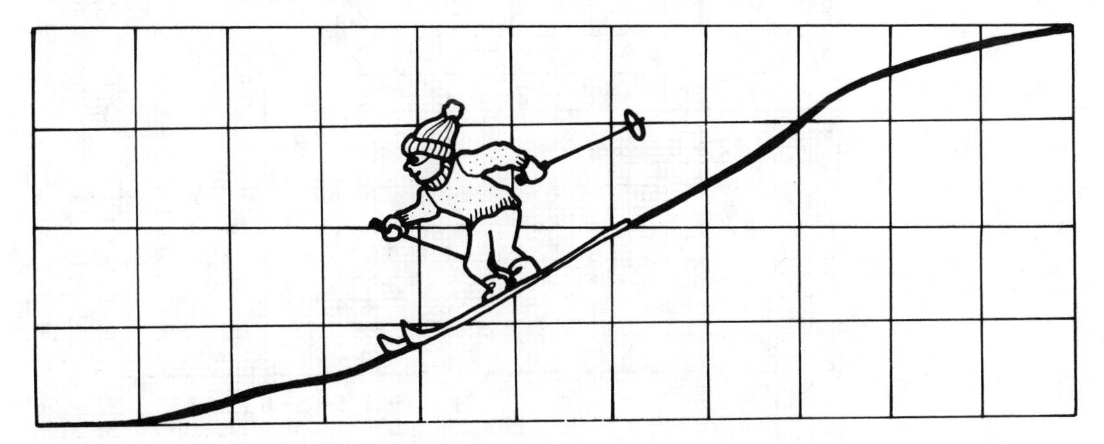

As you already know, graphs can be used to find out—without doing any calculations—desired quantities of a variable by simply reading them off the graph.

Furthermore, by looking at the shape of the graph line of two paired variables, one can tell whether the relationship is direct or inverse. If direct, a look will also tell whether it is a direct proportion.

Also, a graph has a unique application that makes it especially useful for real-world data and calculations. It can be used to quickly and easily smooth out the variations in data that occur due to collecting the data under less-than-perfect conditions.

Even that does not finish the list of what a graph can tell us. By linking up the ideas about proportions presented earlier in this book, a graph can be used to determine the algebraic relationship between its variables.

This chapter will be devoted to a presentation of these uses of a graph.

7.2 REVIEW OF READING GRAPHS

Figure 1 is a number graph (not quantities since no units are given). Use Figure 1 to answer these questions:

1. The number pair of data point A is: $x =$ ______, $y =$ ______.

2. The number pair of data point C is: $x =$ ______, $y =$ ______.

   Another term for "number pair" is "coordinates;" x and y each have a coordinate.

3. The coordinates of data point B are: $x =$ ______, $y =$ ______.

4. Suppose that Figure 1 is redrawn so that the x-axis is re-scaled as shown in Figure 2. (Note that the scale on the x-axis in Figure 2 differs from that in Figure 1.)

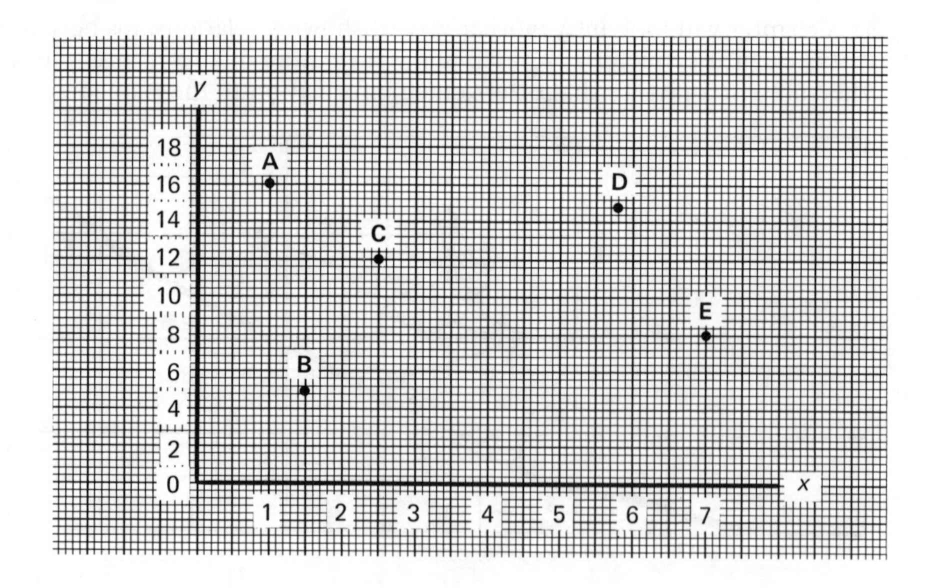

FIGURE 1

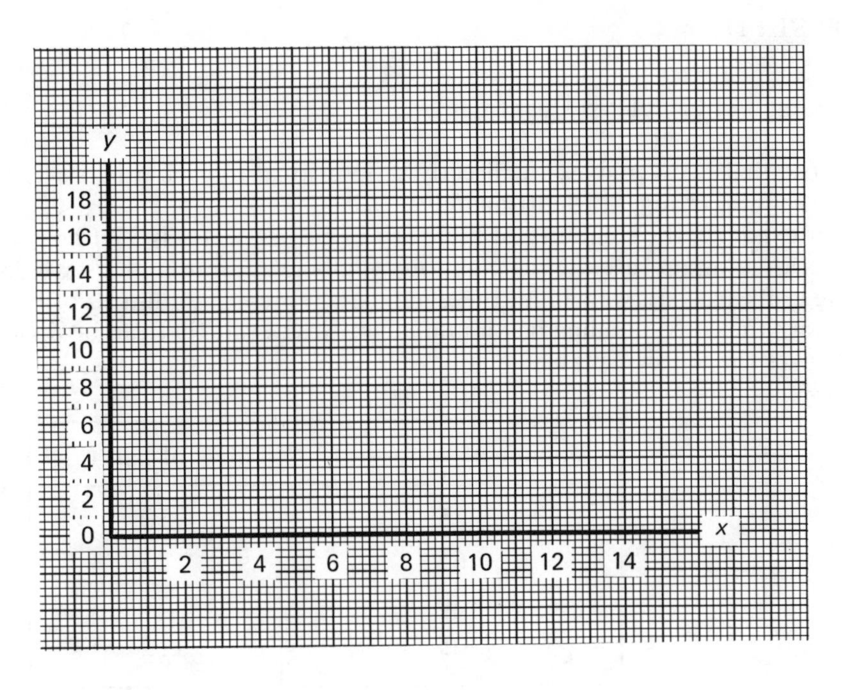

FIGURE 2

a. Where will point E go this time, to the same geographic place in the grid as in Figure 1 or at the new place for $y = 8$, $x = 7$?

b. Does the appearance of the graph change, and, if so, how?

c. Do the coordinates change for any point on the graph? Why?

We see that the appearance of a graph depends not only on the number of pairs that are being graphed but on the scales chosen for the axes. This is an important idea to be kept in mind while reading this chapter and will be discussed further at the end of it. The number pairs shown on the graph change in position when the scale changes but do not change in value.

5. Construct a table of data for the first graph. Such tables are the usual way to present the data from which any graph of real measurements is constructed.

7.3 LINE OF BEST FIT

A. Graph of Calculated Quantities

Suppose that two variables are directly proportional to each other; every time one changes, the other also changes proportionately.

For example, the solubility of table salt in boiling water is 39.12 g per 100 mL water. This is an invariant rate. Let us pick four numbers at random, let's say 25, 65, 130, 305. Using units analysis, we will let these numbers represent millilitres of boiling water and calculate the corresponding maximum quantities of salt that can be dissolved in the water. This gives the following table:

Volume of Water (mL)	Mass of Salt (g)
25	10
65	25
130	51
305	119

The data is plotted as shown in Figure 3 with volume on the x-axis and mass on the y-axis. A line is drawn which, in this case, passes smoothly through all of the data points in a straight line.

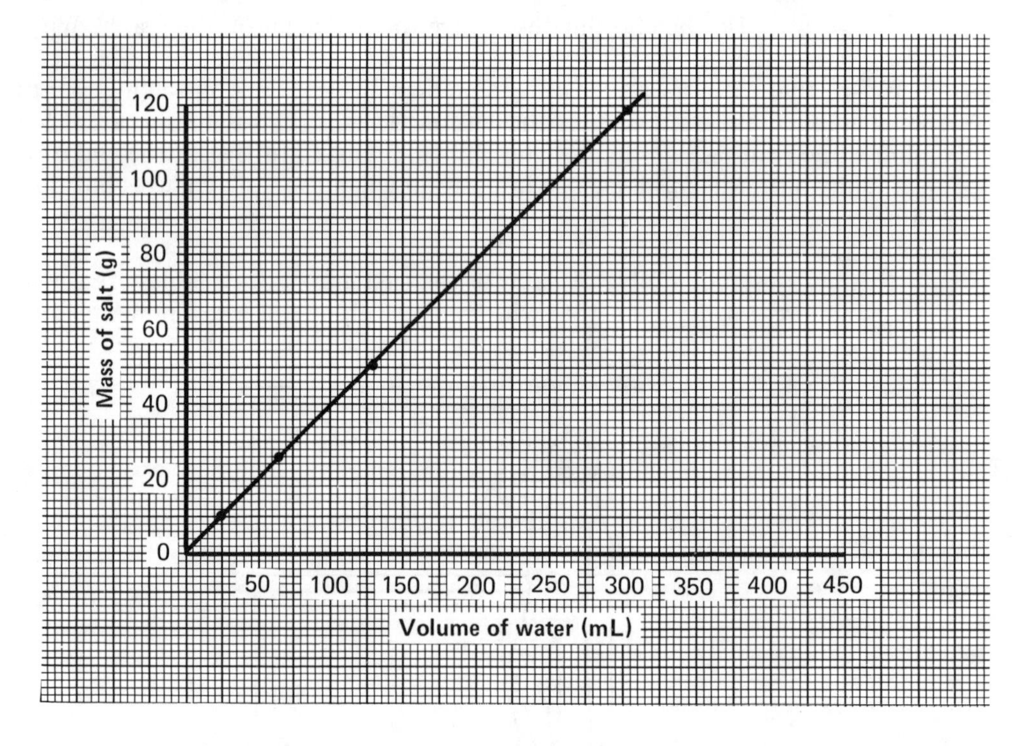

FIGURE 3

This plotted line now enables us to read the coordinates for any selected point on the line. Each point tells us the largest quantity of salt in grams that can be dissolved by a selected volume of boiling water in millilitres. It also tells us the least volume in millilitres of boiling water needed to fully dissolve a selected mass of salt in grams.

For example, Figure 3 shows that the volume of water needed to dissolve 40, 60, and 80 g of salt read off from the graph, is about 103, 154, and 204 mL respectively. Each large scale marking on the x-axis counts for 50 mL water, with the ten small scale markings within this interval counting for 5 mL each. Each large scale marking on the y-axis counts for 20 g each with the ten small scale markings within this interval counting for 2 g each.

1. From the graph of Figure 3, what is the solubility of salt in 250 mL of boiling water?

2. Using Figure 3, how much water is needed to dissolve 46 g of salt?

3. Given the solubility of 39.12 g salt per 100 mL of boiling water, *calculate*:

   a. Salt solubility in 250 mL boiling water.

   b. Quantity of water needed to dissolve 46 g of salt.

   c. How do the graphical results compare with the calculated ones? Account for any differences.

Notice that the data appears in the upper right quadrant of the axes as shown below:

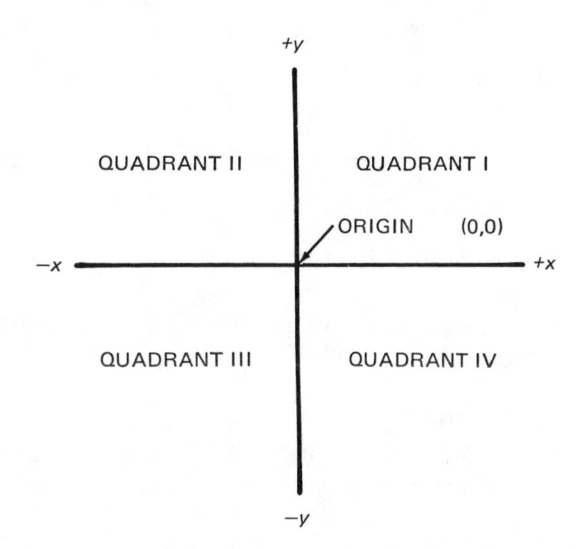

In science, most measurements use a scale that starts at zero; hence, there are no negative quantities for such measurements. The most common exception to this is the Celsius scale which has its zero well above the lowest temperature possible and thus allows negative temperature readings.

Most graphs in science use only Quadrant I. In this book, the graphs will likewise be confined largely to Quadrant I.

Solve the following problems:

4. a. Plot a graph and draw the line to fit the numerical values of the following table.

x	y
2	4
5	25
6	36
8	64

   b. If $x = 2.5$, what is y?

   c. If $y = 50$, what is x?

5. a. Plot a graph and draw the line that fits the numerical values in the following table.

x	y
0	-10
10	10
15	20
22	34

   b. If $x = 2$, what is y?

   c. If $y = 17$, what is x?

6. a. Plot a graph and draw the line that fits the numerical values in the following table.

x	y
10	40
20	20
26.7	15
80	5

   b. If $x = 25$, what is y?

   c. If $y = 25$, what is x?

B. Ideal versus Real

The data for the graph in Figure 3 were calculated from a rate that was given as invariant. Let us consider what happens when all the data is measured rather than calculated.

When taking measurements, there are many influences that can cause them to vary from what would be expected under ideal conditions. Suppose we have 10,000 Mexican jumping beans on a table and wish to count the number of jumps per one second, two seconds, and so on. Everyone in the class takes the count; everyone comes up with different numbers. For one thing, not the same number of beans jump each second. Also, is a jump any motion, motion off the table, or a motion that turns the bean? Even when this is decided, it is hard to be sure at which second some of the beans started to move, and easy to miss one bean's motion while spotting another's.

This kind of difficulty occurs with all measurements, no matter how superb the instruments and methods are. Suppose you wish to measure the line drawn below.

If you observe it under a magnifying glass, you will find that the ends are ragged. Where does the line actually begin and end? Under high magnification, your ruler edge also turns out to be ragged. A difference in measurement of one-ten-thousandth of the width of this page can cause a Mars probe to miss Mars.

There is variation built into all measurements. Depending upon the objects to be measured, the method and the instrument, all measurements tend to vary from the ideal result.

As a classroom experiment on this, your teacher will hand you one or more objects to measure such as cubes of sugar. You will be asked to plot your measurement on a graph on the blackboard. Everyone in the class will add their data to it. Can a single straight line be drawn through all the data points?

It is most unlikely that the data will all fall on a straight line. Ideally it should, but your instrument is not perfect, and the cubes of sugar or other objects you weigh may not be perfectly uniform. However, while some readings are too low, others may be too high. If the variations occur by chance—as much in one direction as in the other—then we can assume that half the data points fall above the ideal data and half below. Hence, if a line is drawn, for example, between all of the points as shown, the line may well be very close to where the ideal data would be plotted.

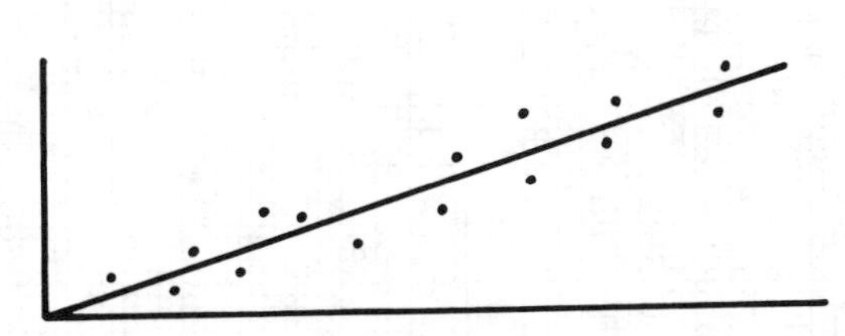

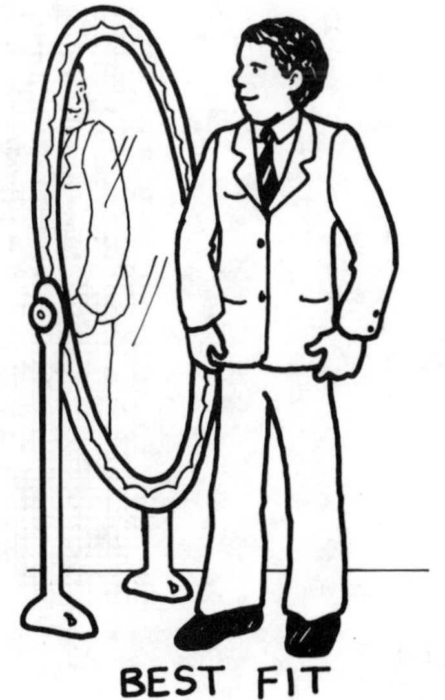

Such a line is called *a line of best fit.* The points on it are probably closer to the ideal than most of the measured data.

As another example, suppose that instead of assuming that the solubility of salt in boiling water is an invariant rate as in Section 7.3A, we choose instead to measure the solubility. To check this, it will be necessary to find out how much salt dissolves in several different quantities of boiling water. This may not be too easy. For one thing, how do you know that more salt cannot be dissolved in a certain quantity of water than is already put in? One way to find out is by adding *too much*; if the excess doesn't dissolve after much stirring, the solution is then holding as much salt as it can. Now you must filter off the excess salt while keeping the solution at boiling. Alas, some of the water may boil out. Furthermore, unless your water is quite pure, it may be boiling at a temperature higher than 100°C, when the salt is more soluble. After you filter off the salt, you will have to figure out how much filtered boiled salt-water you have and boil it until dry, recover all the salt even though some may spatter, and weigh it. Just to make sure, repeat the whole thing. What are your chances of twice getting the same results for solubility?

In an actual measurement of salt solubility, the following results were obtained.

Volume of Water (mL)	Mass of Salt (g)
50	10
60	24
95	50
233	80
282	122

These are plotted in Figure 4. The line connecting all the points is not a single smooth continuous line but a jagged one. Such a line shows a rate that alters from point to point (to be discussed later); the rate does not appear to be invariant. However, Figure 5 shows a graph where these chance variations have been smoothed by the line of best fit. In Figure 5,

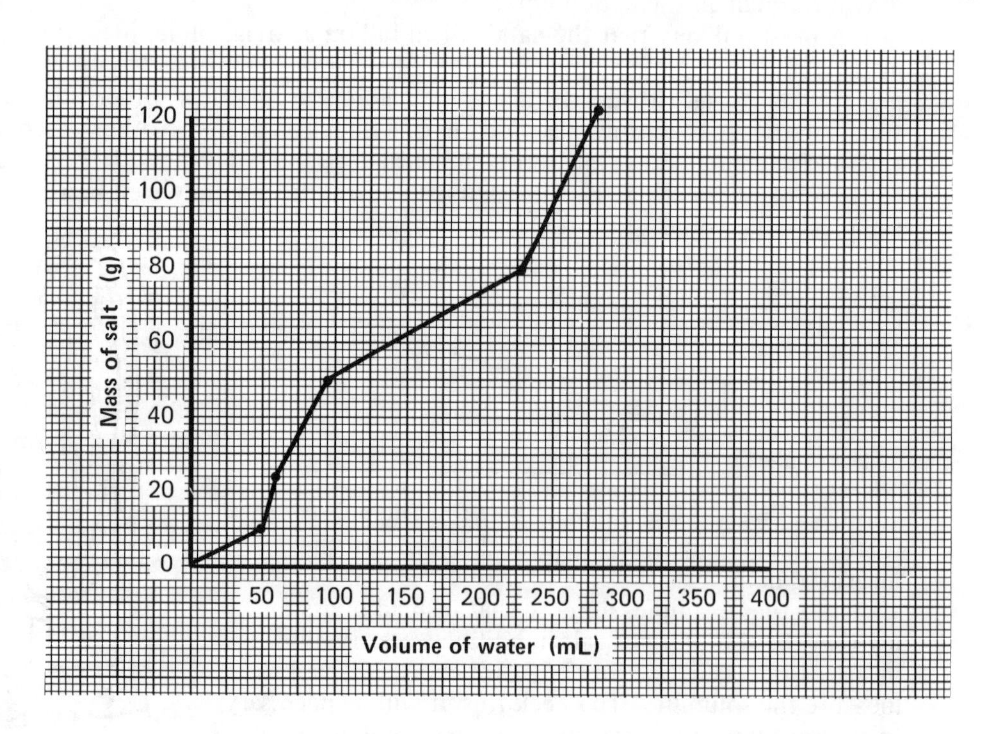

FIGURE 4

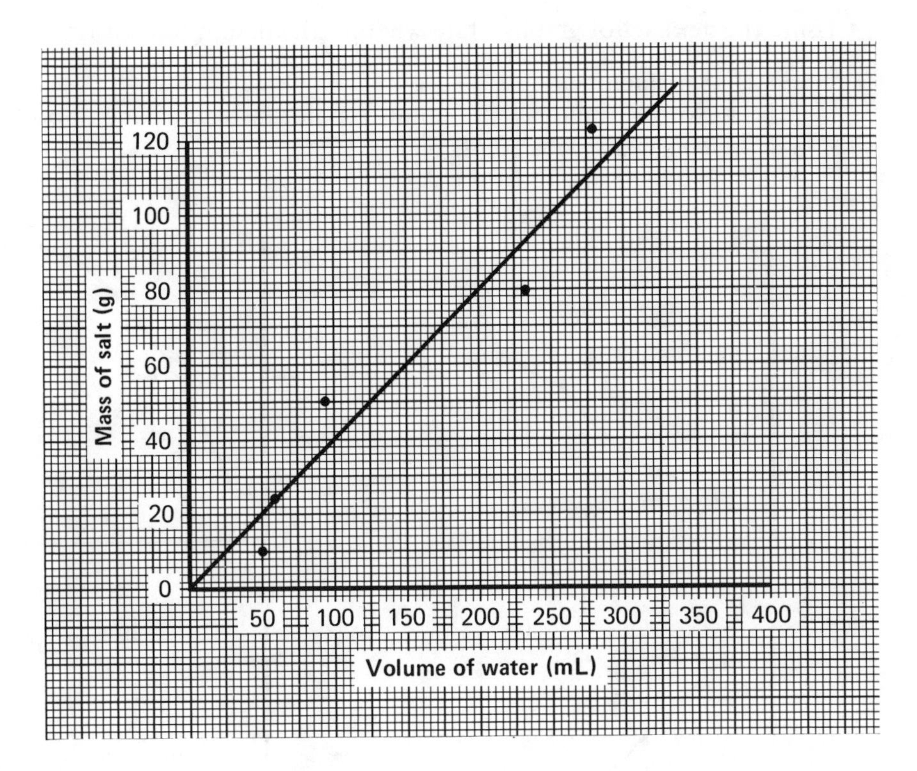

FIGURE 5

the line has been drawn to pass between all the data as closely to them as possible, judged by the eye. The graph in Figure 5 closely resembles that for the ideal data shown in Figure 3. If the variations from the line of best fit are random (a matter of "luck"), then the line has a fairly good chance of resembling the ideal line.

The practice of drawing a line of best fit is much used in experimental studies. Here are some examples of unmarked graphs with data points and lines of best fit.

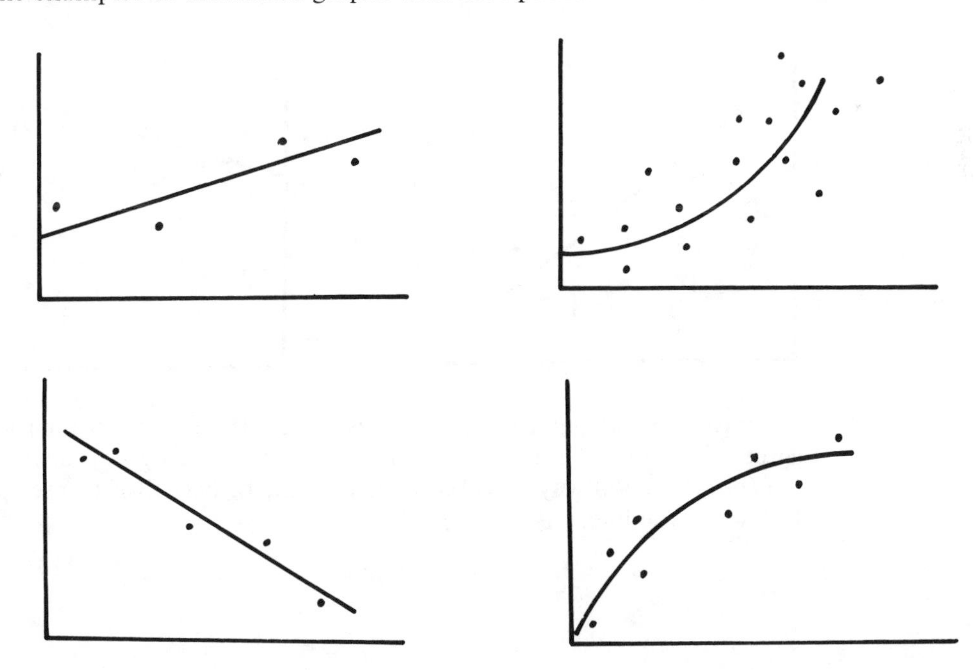

Using the next set of graphs, state whether the line of best fit is:

a. Satisfactory.
b. Not continuous (data point-to-point instead of one smooth line).
c. Too many data points on one side.
d. Wrong shape to fit the data.

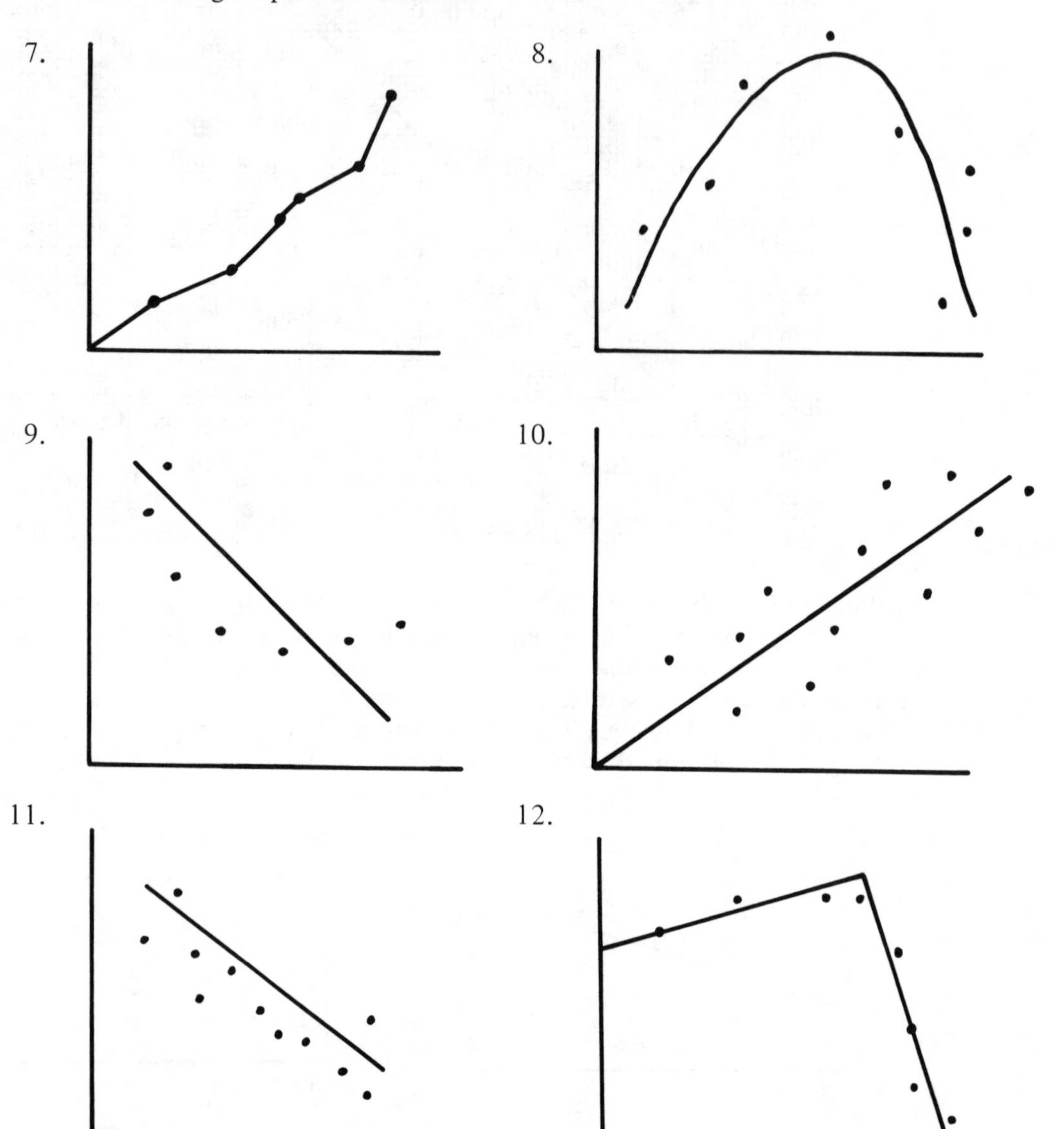

13. Copy the following graphs. Draw lines of best fit. There is some room for disagreement about the drawing of the line, since these are not ideal quantities. (*Hint:* A good way to visually select a line of best fit is to use a black thread to lay out the line.)

a.

b.

c.

d.

e.

f.

g.

h.

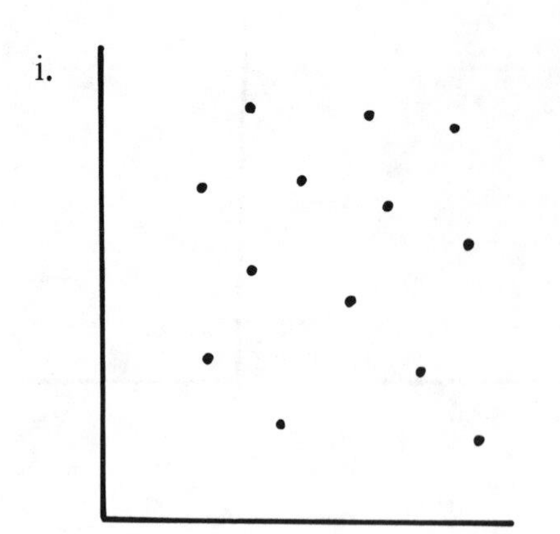

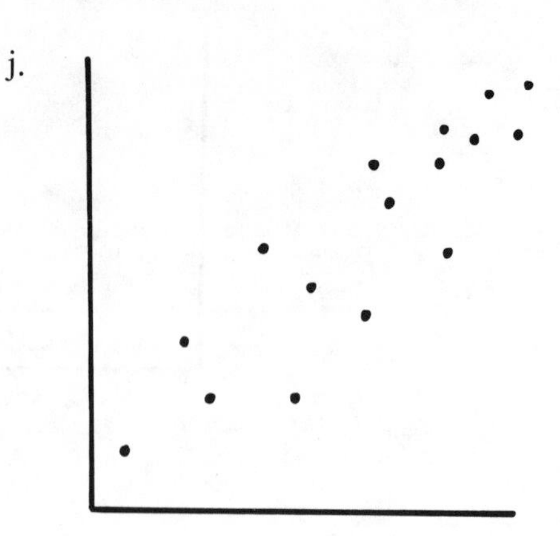

7.4 READING A DIRECT OR INVERSE RELATIONSHIP

Each graph is actually a picture. It pictures the kind of relationship the variables have. Just one look at it tells you whether the relationship is direct or inverse. As we shall see later, it also tells you whether a relationship is proportional, exponential, additive, or subtractive.

At this point, let us see how a graph tells whether a relationship is direct or inverse. If a relationship is direct, as x increases, y increases. That means that wherever the line of best fit starts, the line must then swoop up to the right.

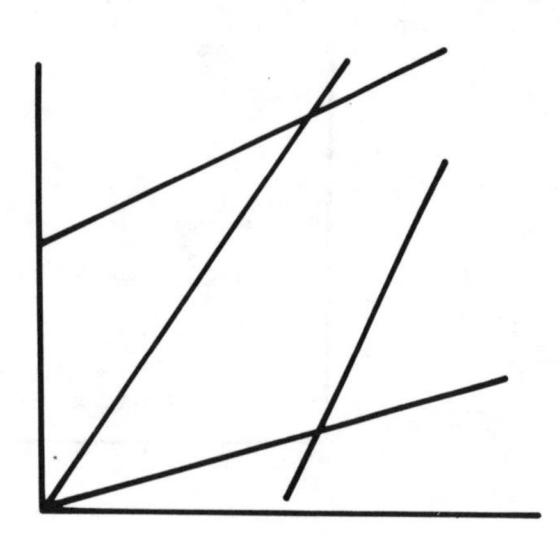

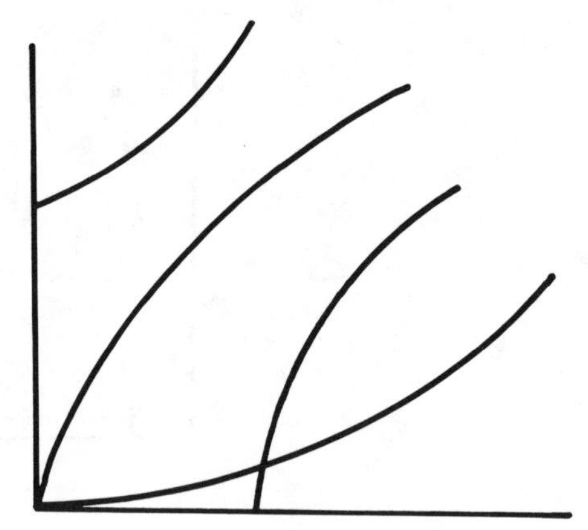

The above all show direct relations.

Conversely, if the relationship is inverse, then the line must move downwards from left to right. The following all show inverse relations.

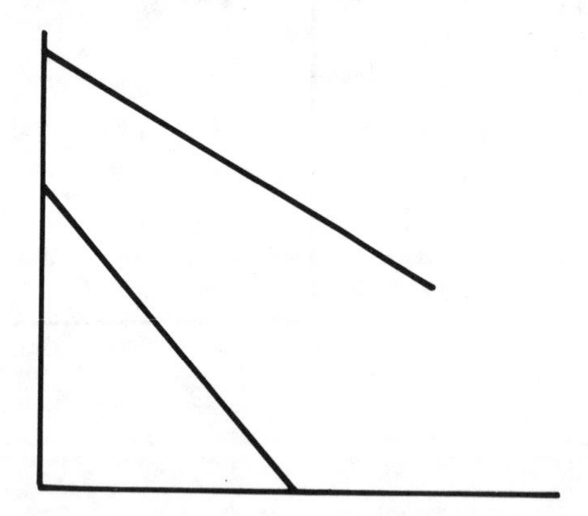

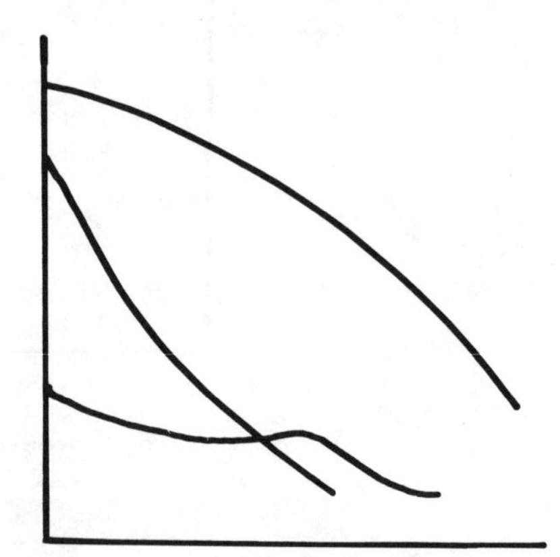

1. Illustrate with the use of the salt solubility graph (Figure 3) in Section 7.3A that a direct relation shows a line rising toward the right.

2. Illustrate with the use of the graph in Section 7.3A, Problem 6, that an inverse relation shows a line sloping down from left to right.

3. State which of the graphs show direct relations, which show inverse relations, and which show neither or both in Section 7.3B, Problems 7 through 13.

4. Draw 3 sets of data points that indicate in different ways that y is not related to x.
   (*Hint:* See Problem 13 in Section 7.3B.)

7.5 SOME GRAPHING SKILLS

In order to construct a graph, certain rules must be followed to make sure that the picture honestly tells what the data says. These rules will be discussed next.

A. Dependent and Independent Variables

Customarily, the independent variable is graphed on the horizontal axis while the dependent one is written on the vertical axis.

The independent one is the one that is manipulated while the dependent one responds.

The horizontal axis is also called the abscissa or, sometimes, the x-axis. The vertical axis is also known as the ordinate or, sometimes, the y-axis.

In the following, determine which variable should be charted on the x-axis and which on the y-axis.

1. A biologist plants varying numbers of seeds on a 10 ft. × 10 ft. plot and then counts the number of seedlings that come up.

2. A hose has a nozzle whose opening can be adjusted. As the size of the opening is adjusted, the distance reached by the stream of water is measured.

3. Candles of different diameter are burned. The change in height is measured with the passage of time.

4. The volume and mass of each of a batch of rocks are measured.

B. Uniform Scale

When an axis is marked off into line segments, each segment must be equal in value. The scale below is wrong, because the interval between 0 and 5 is worth 5 but the interval between 10 and 20 is 10. The scale is not uniform and will not present an honest picture of change.

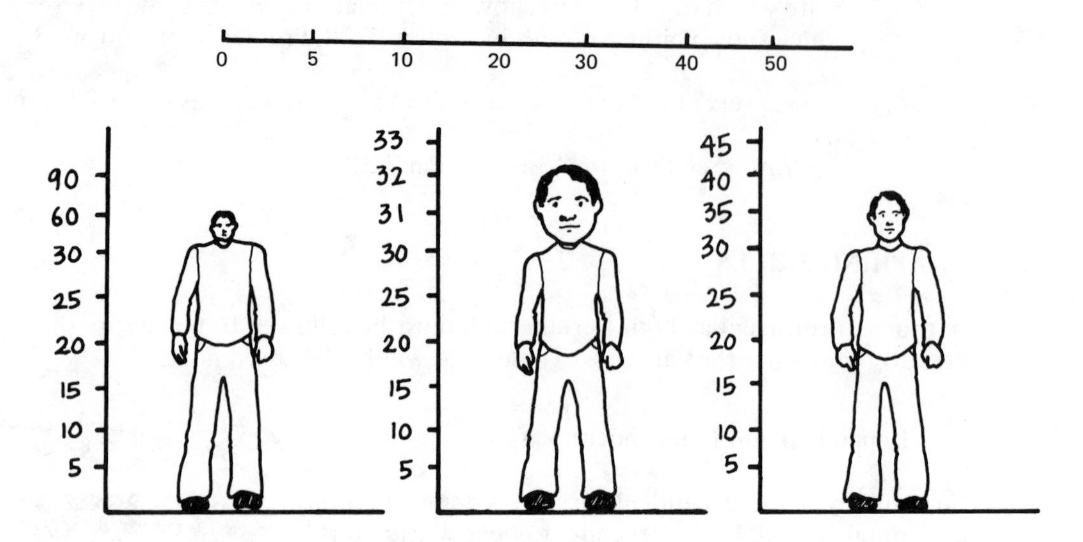

To find out if the scale markings are uniform, subtract each from the one following it. The difference should be the same each time (except for logarithmic scales, not covered here).

Which of the following is incorrect because the scale is non-uniform?

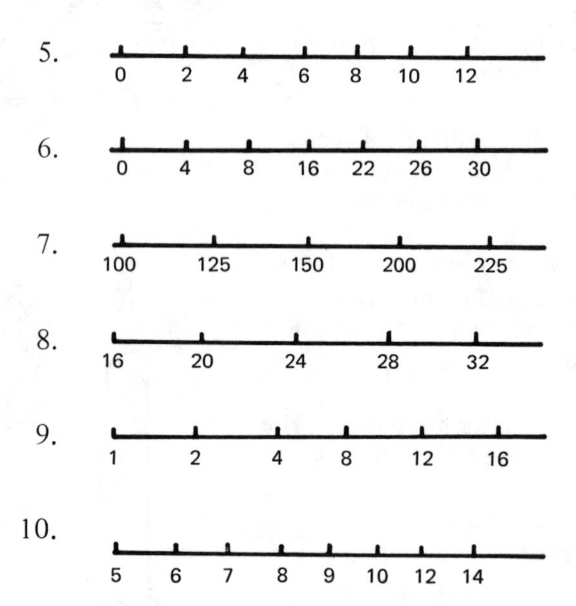

C. Make the Picture Fit the Frame

Given a sheet or half-sheet of paper, the graph should not be scaled so that all the data points fit into one small portion of it. Use up as much of the available space as possible. The bigger the graph is, the easier it will be to read it. Don't use a big sheet of paper to draw a graph with a little line of best fit.

A graph is a mathematical portrait. A portrait which is squeezed into one portion of the paper is hard to see.

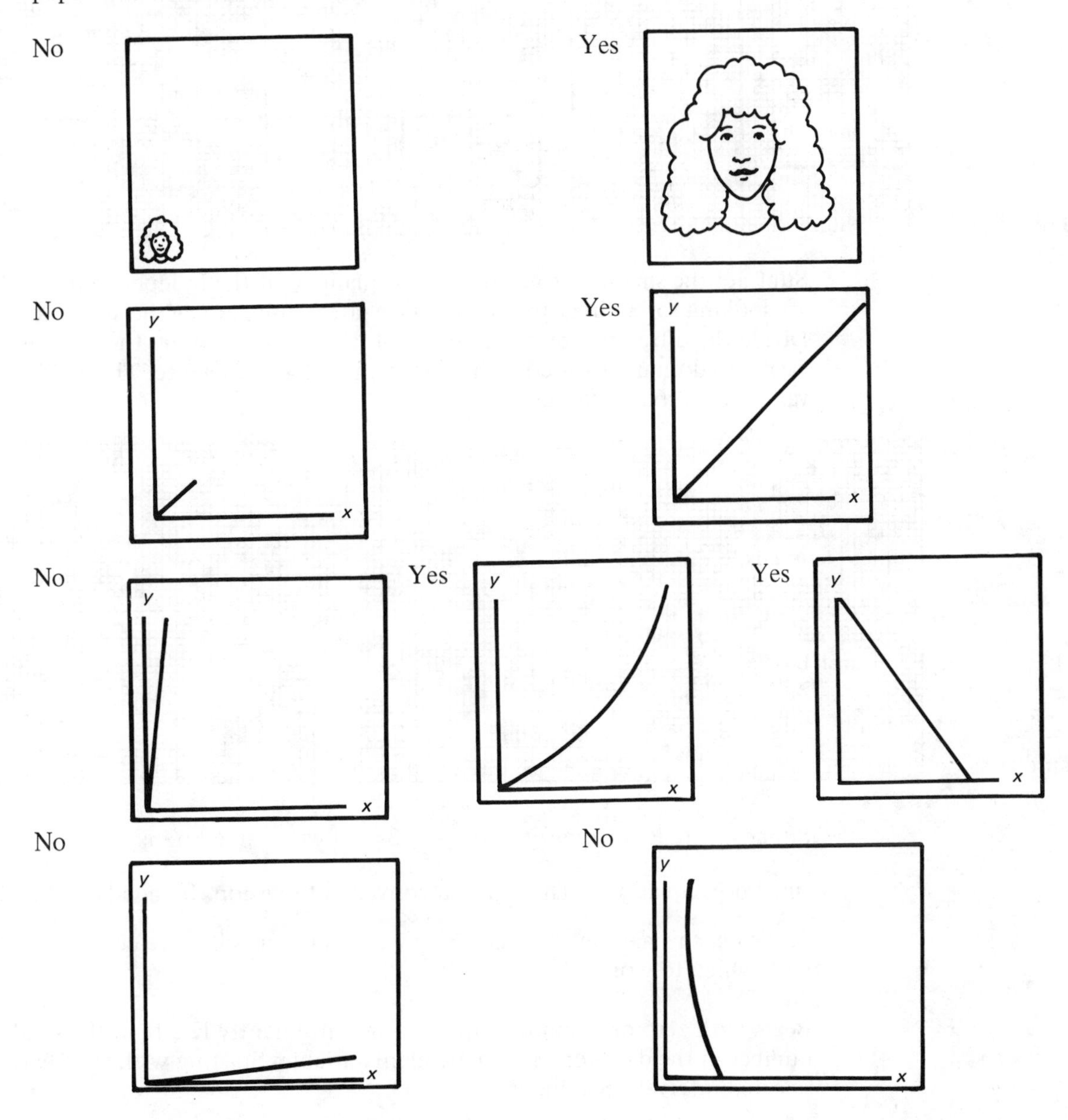

To mark a useful scale on an axis, first note that the graph paper is divided into large boxes which in turn are divided into small boxes.

1. Draw the x-axis about 1 to $1\frac{1}{2}$ inches from the bottom along a large box line running the width of the page. Leave a margin at the left of about 1 inch per 8 inches of width. Count off the number of large boxes. See Figure 6a.

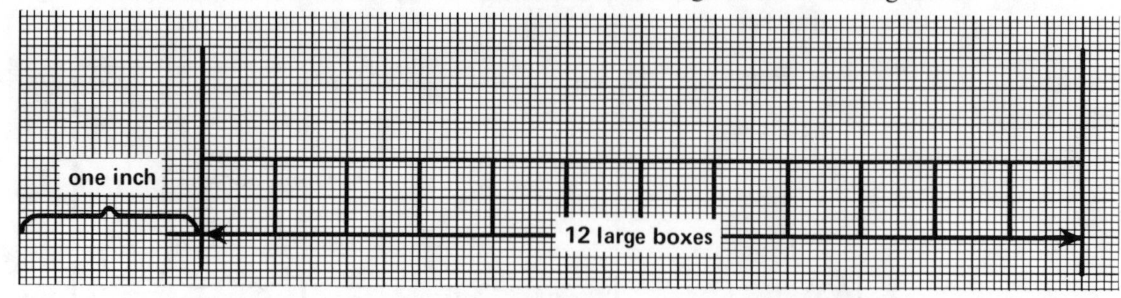

FIGURE 6a

2. Subtract the smallest from the largest quantity of the independent variable. If you are looking for a direct proportion (see later), use zero as the smallest quantity. Divide the difference by the number of large boxes. Round the number off to an easily used number for counting off such as 2, 4, 5, 10, etc. This will be the interval on the x-axis. See Figure 6b.

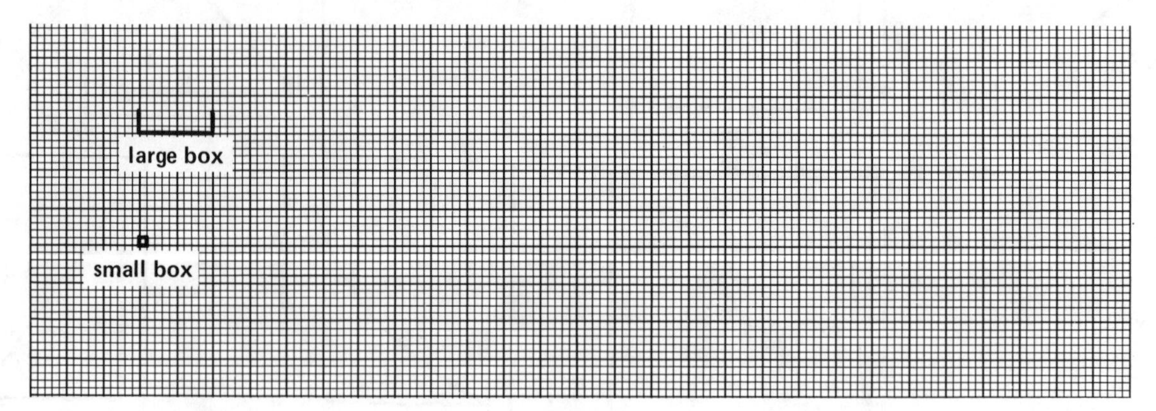

FIGURE 6b

Suppose x ranges from 0 to 40; $\frac{40}{12} = 3\frac{1}{3}$. If each large box is set at 4, then each small box equals 0.4. This is not a convenient division. If each large box is set at 5, then each small one equals 0.5 or $\frac{1}{2}$. This is a much more convenient interval with which to work.

3. Begin with zero or a number equal to or conveniently less than the smallest number in the data for the independent variable. Starting with this, mark off the selected intervals. See Figure 6c.

4. Follow a similar procedure for the y-axis. It is not necessary to have both axes plotted to the same scale.

0 5 10 15 20 25 30 35 40 45

FIGURE 6c

For the following table, select the axis with the best labeling, assuming that each axis represents the width of the page. Explain your choice.

x	y
5	136
8	126
11	114
16	100
24	96

For the x-axis:

a. 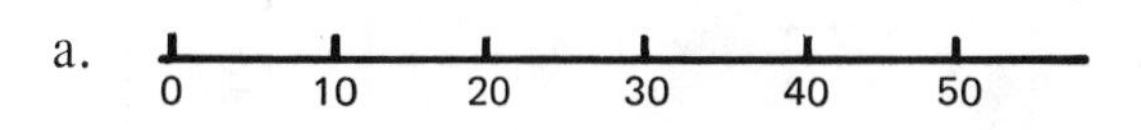

b. 0 2 4 6 8 10

c. 0 5 10 15 20 25

d. 0 6 9 15 18 24

For the y-axis:

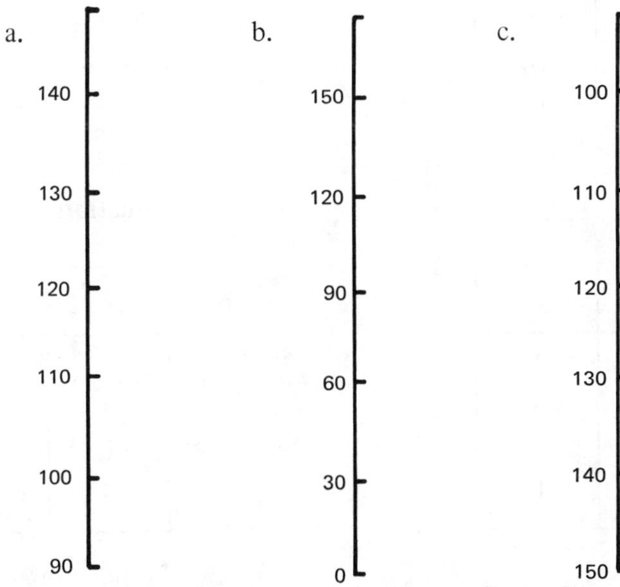

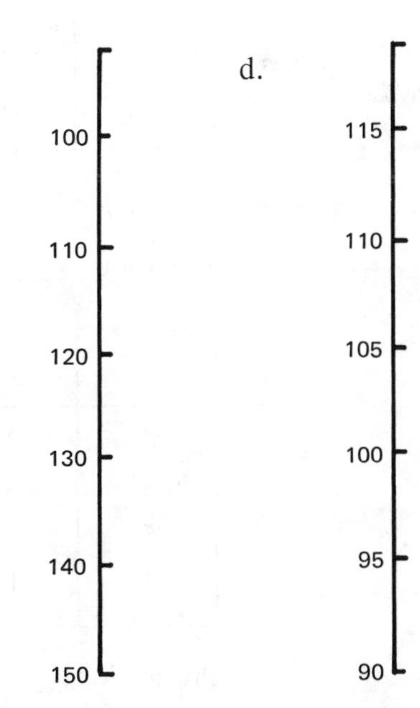

7.6 GRAPHS OF DIRECT RELATIONS

For simplicity, we will work in this section with numbers rather than quantities, and will use integers.

A. Constructing Graphs of Direct Proportions

In each of the following series of tables, determine whether the relationship is a direct proportion or some other direct relation. If the relation is a direct proportion, determine the equation. Then, graph all of the direct proportions on the same graph so that the graph takes up most of an 8½ × 11″ sheet of graph paper. Similarly, graph all of the other relationships on a second sheet of graph paper. In all graphs, extend the line to intersect the y-axis.

1.

x	y
1	5
2	10
3	15
4	20

Proportion? ________________

Equation ________________

2.

x	y
3	5
5	7
2	4
6	8

Proportion? ________________

Equation ________________

3.

x	y
1	1
2	4
3	9
4	16

Proportion? ________________

Equation ________________

4.

x	y
3	9
4	12
2	6
5	15

Proportion? ________________

Equation ________________

5.

x	y
6	3
3	0
5	2
4	1

Proportion? ________________

Equation ________________

6.

x	y
1	6
3	18
2	12
4	24

Proportion? ________________

Equation ________________

7.

x	y
2	4
3	5
5	6
7	7

Proportion? ______________

Equation ______________

8.

x	y
2	8
1	4
3	9
5	25

Proportion? ______________

Equation ______________

9.

x	y
1	5
2	6
3	8
4	11

Proportion? ______________

Equation ______________

10. Compare the graphs of direct proportions with those that are direct but not proportions. How can you tell from the graph when the relationship is a direct proportion?

11. Which of the next group of graphs show variables related by a direct proportion?

a. 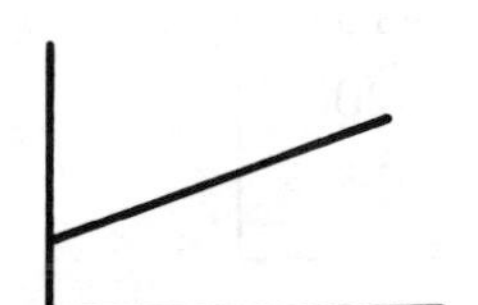

b. 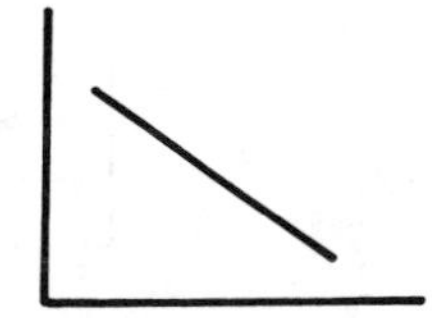

c. 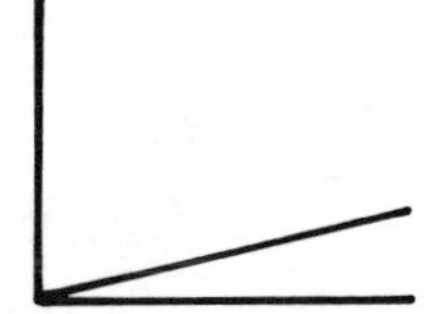

d. 

e. 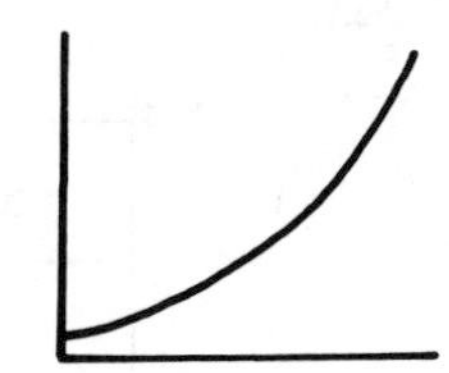

f. 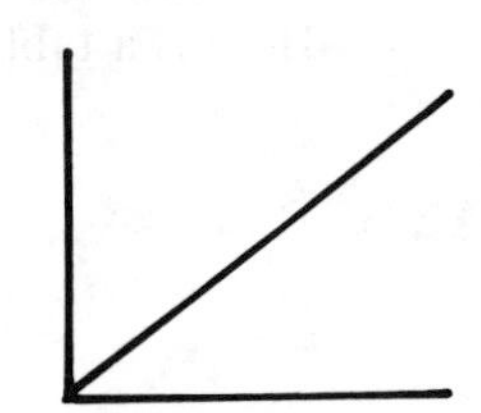

B. Two Criteria for Graphs of a Direct Proportion

The section preceding this showed that a graph of direct proportion always shows *a straight line passing through the origin.*

We already know that the general equation for a direct proportion is $y = kx$. Thus, $y = kx$ must be the general equation for any straight line passing through the origin. Let us see how this agrees with the two requirements for the graph of a direct proportion, namely, that the line must be straight and pass through the origin.

1. **The First Criterion: Must Go Through Origin**—Answer the following questions.

12. If a line passes through the origin, then when $x = 0$, what is the value of y?

13. In the equation, $y = kx$, if $x = 0$, what is the value of y? Does this agree with the answer to Problem 12?

14. If you can buy 5 electric light bulbs for \$2, how many can you buy for \$0? Does this agree with your answers to Problems 12 and 13?

Evidently, when two variables are directly proportional to each other, if one shrinks to zero, so must the other. Hence, a graph of the relationship must show this. Thus, a graph of a direct proportion must go through an origin.

2. **The Second Criterion: Must Be a Straight Line**—Now let us consider why the line of best fit of a direct proportion must be a straight line.

A direct proportion is characterized by an invariant rate. Given the invariant rate $y/x = k$, for every one quantity of x, there is always the same quantity of y. For two quantities of x, there is twice as much y, and so on.

Suppose $y = 5x$. The following table may be constructed.

x	y
1	5
2	10
3	15
4	20
5	25

Every time that x increases by 1, y must increase by 5.

Here is a table for $y = \frac{1}{2}x$.

x	y
1	½
2	1
3	1½
4	2
5	2½

Every time that x increases by 1, y must increase by ½.

It is this steady rate of increase that causes a straight line. If x increases by one, y increases by, let's say, ½, as in the equation $y = ½x$.

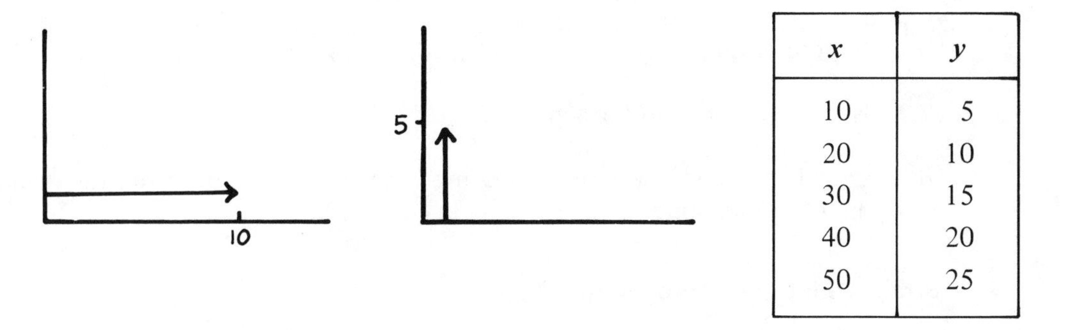

x	y
10	5
20	10
30	15
40	20
50	25

Together, they increase by

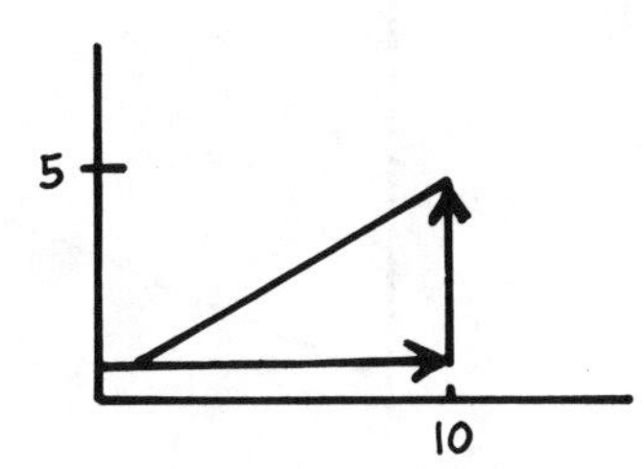

This continues along, resulting in a straight line, as shown in Figure 7.

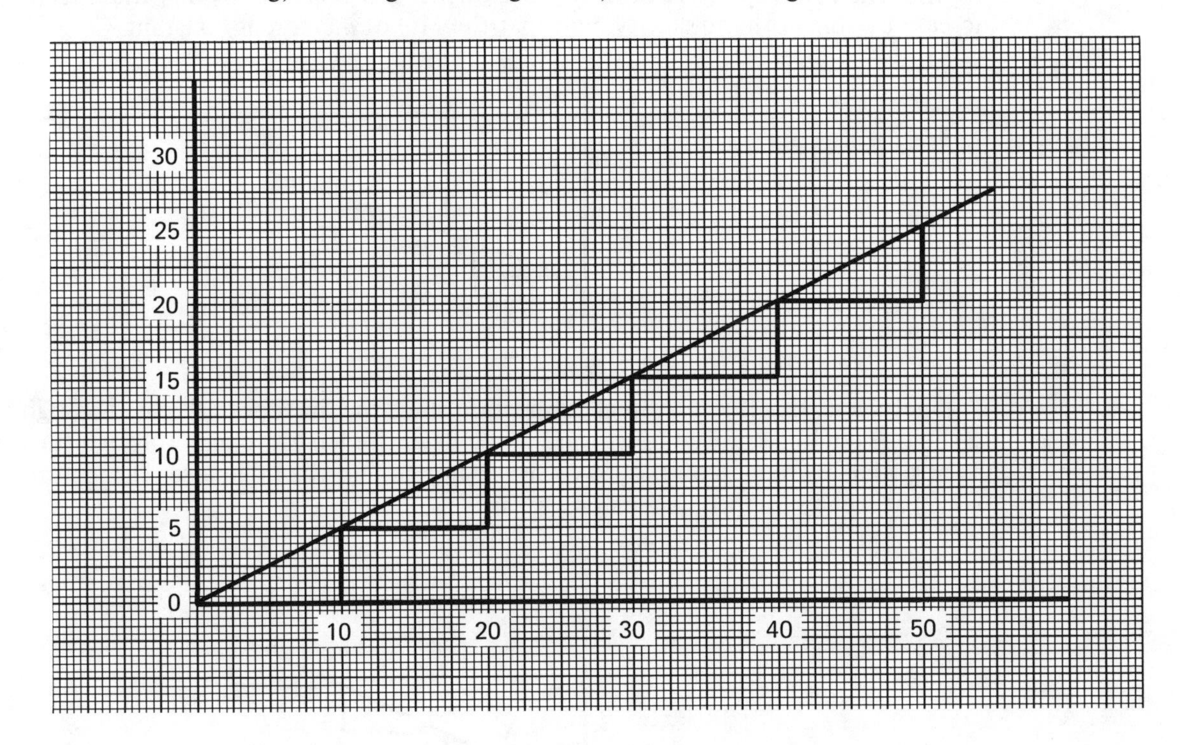

FIGURE 7

Regraph Figure 7, using double the number of boxes for each scale marking on the y-axis.

15. Does your graph show a direct proportion? How do you know?

16. In your graph, when x increases by 10, does y still increase by 5?

17. Is y/x equal to ½ in both graphs?

18. Can you suggest a way of calculating k, the invariant rate, from the graph of a direct proportion?

C. Straight Line, Not Through the Origin

What kind of a relationship exists for a graph such as this?

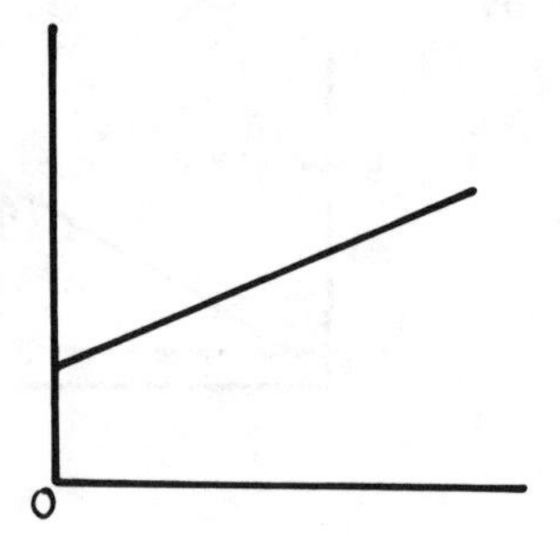

We see that, since this is a straight line, whenever x increases by any quantity, y must increase by some other quantity, and that the rate of *increase* is constant.

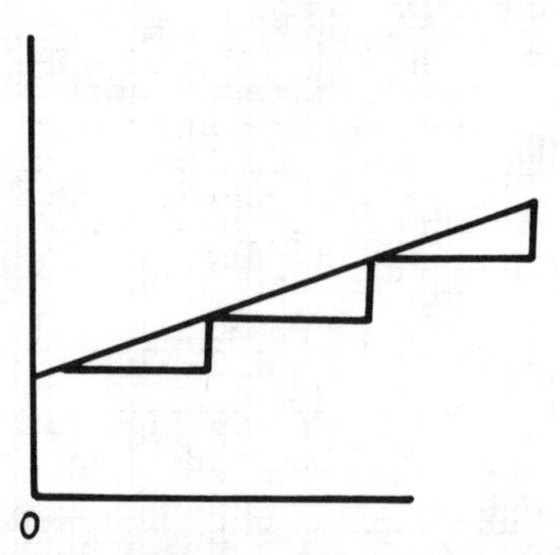

Nonetheless, this is not a direct proportion since the line does not pass through the origin.

The line could be moved downward along its entire length to pass through the origin by altering what is graphed on the y-axis. First, notice that the line passes through the y-axis ($x = 0$) at some value we shall call b.

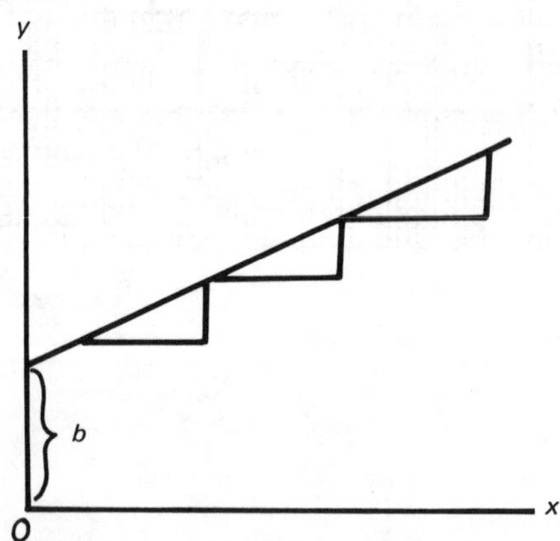

If the quantity b is subtracted from all of the y quantities and $y - b$ plotted against x, then the entire line is lowered on the graph by the amount of b. At $x = 0$, $y - b$ equals zero.

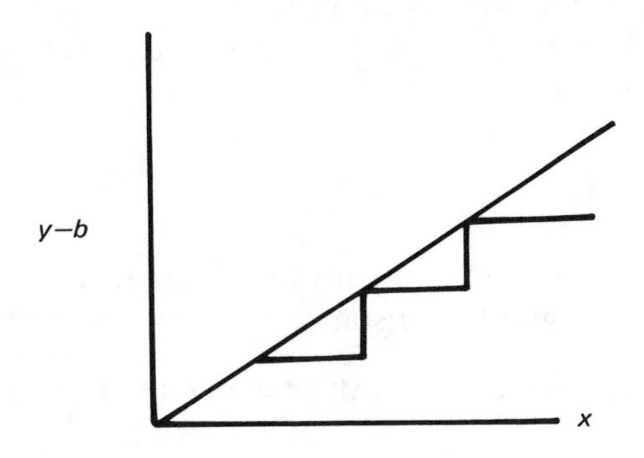

Now the graph shows a direct proportion between $y - b$ and x. Thus,

$$y - b = kx$$

or $$y = kx + b.$$

See Section 5.7B for a discussion of this from a different viewpoint.
Answer the following.

19. Given the equation $y = kx - b$, which is related to x by a direct proportion, y, $y - b$, or $y + b$?

20. Graph the equation $y = 5x - 2$ selecting your own numbers for the independent variable x.

21. Graph the equation $y = 3x + 6$, selecting your own number for the independent variable x.

22. Graph the equation $°F = \frac{9}{5}°C + 32°$.

23. For each equation in Problems 20 to 22, what is directly proportional to the independent variable? How does the graph show this?

24. a. Rewrite each of the equations of Problems 20 to 22 to show a constant rate and state the value of the constant.

    b. Suggest a way to calculate the invariant rate from the graph. (*Note:* The invariant rate is called the *slope* in mathematics.)

25. Here's one last point that you can figure out from the graph. For each equation in Problems 20 to 22, where does the number appear that has to be subtracted from y to make it proportional to x? It is called the *y intercept.*

7.7 GRAPHS CAN TELL RATES

So far, we have seen that the equation for graphs of $y = kx$ and $y = mx + b$ (an m is used instead of k to help distinguish between the equations) can be identified just by looking at the graphs.

The graph that shows a straight line going through the origin is the graph of $y = kx$. If the graph shows a straight line that crosses the y-axis above or below the origin, the equation is $y = mx + b$ where b is the intercept on the y-axis.

Given such graphs, the information needed to calculate the invariant rate can readily be obtained from the graph.

A. The Case of $y = kx$

Let us consider how to calculate the invariant rate of y to x from a graph of $y = kx$, or of distance to time from a graph of distance and time or of cost per quantity of items from a graph of cost and quantity of items, or of Z per from a graph of Z per . They all use the same idea.

For the graph of y versus x given the equation $y = kx$, all that is needed is to select any point on the line of best fit, read off x and y, and calculate y/x, since $y/x = k$ = the invariant rate.

Let us try this by calculating the invariant rate for solubility of salt in boiling water from the graph of Figure 5 in Section 7.3B.

First, arbitrarily pick any coordinate of x such as 300 mL water. It is a good idea to pick a large rather than a small quantity. When volume of water = 300 mL, what is the mass of dissolved salt? We read it as about 117 g salt and use it to calculate the unitary rate:

$$\frac{117 \text{ g salt}}{300 \text{ mL water}} = \frac{0.390 \text{ g salt}}{\text{mL water}} = k = \text{the invariant rate.}$$

The equation for this graph is:

$S = kW$ where $k = 0.39$ g salt/mL water.

Let S = mass salt, W = volume water, k = constant.

Do the following problems:

1. Select two other points on the line of best fit in Figure 5, and calculate the invariant rate for each. Does the rate change from point to point of a straight line through the origin?

2. a. Calculate the invariant rate (units omitted) for Figure 7.

   b. Calculate the invariant rate (units omitted) for the graph of Figure 7 redrawn with a different scaling for x.

   c. Is there any difference between your answers to 2.a. and 2.b.? Explain.

3. a. Calculate the value for k from your graph for Problem 1, Section 7.6A.

   b. From your determination of k, calculate y when $x = 3.2$. How does this compare to the y reading on the graph for $x = 3.2$?

4. a. Calculate the value of the invariant rate from your graph for Problem 4, Section 7.6A.

   b. Calculate y/x from the data table for Problem 4.

c. Compare your answers to 4.a. and 4.b. It is not unusual to find that they differ. Why might this happen?

5. Write the rule for graphically determining the rate for variables related by a straight line going through the origin.

B. The Case of $y = mx + b$

The graph of $y = mx + b$ shows a straight line as does the graph of $y = kx$. When b equals zero, $y = mx + b$ becomes $y = mx$. Whether k or m or any other symbol is used for the constant doesn't matter. Thus, $y = mx$ (or $y = kx$) is the special case of $y = mx + b$ where b equals zero.

We have already examined the special case of $y = kx$, so let us move on to the more general case, $y = mx + b$.

It is often desirable in science to determine the numerical value of the constant m. For example, in the conversion of °F to °C, a useful equation is $°F = \frac{9}{5}°C + 32°$. In this equation, $m = \frac{9}{5}$.

For an equation such as $y = mx + b$, one way to determine m graphically is to plot $y - b$ versus x. Then m can be determined as in the case of $y = kx$.

Ordinarily, the scientist will have already plotted the y and x variables and is not eager to plot a second graph. As you know, graph-making is time consuming and requires exercise of patience. Fortunately, it is possible to calculate m without redrawing the graph.

Before we do this, notice that m is not the rate of y to x in the equation $y = mx + b$. Hence, it has been given another name; it is called the *slope*.

(For the special case of $y = kx$, the slope is k; it is also the invariant rate of y to x and can be readily obtained from the graph of y to x as already discussed.)

To calculate the slope for the graph of any equation of the form $y = mx + b$, first pick two points on the line of best fit. Preferably, these points should be far apart, one at the lower end of the line and one near the upper end. Let us call the y- and x-coordinates at the higher end y_2 and x_2 and at the lower end y_1 and x_1. Then

$$m = \frac{y_2 - y_1}{x_2 - x_1}.$$

This is the general equation for a slope.

(For $y = kx$, the y_1 and x_1 values may be taken as those at the origin. In that case, $y_1 = x_1 = 0$, and $m = y_2/x_2$. This is the same as saying that $k = y/x$.)

As an example, the data below is graphed in Figure 8.

t (seconds)	*D* (miles)
1	0.83
2	1.06
3	1.22
4	1.36
5	1.52

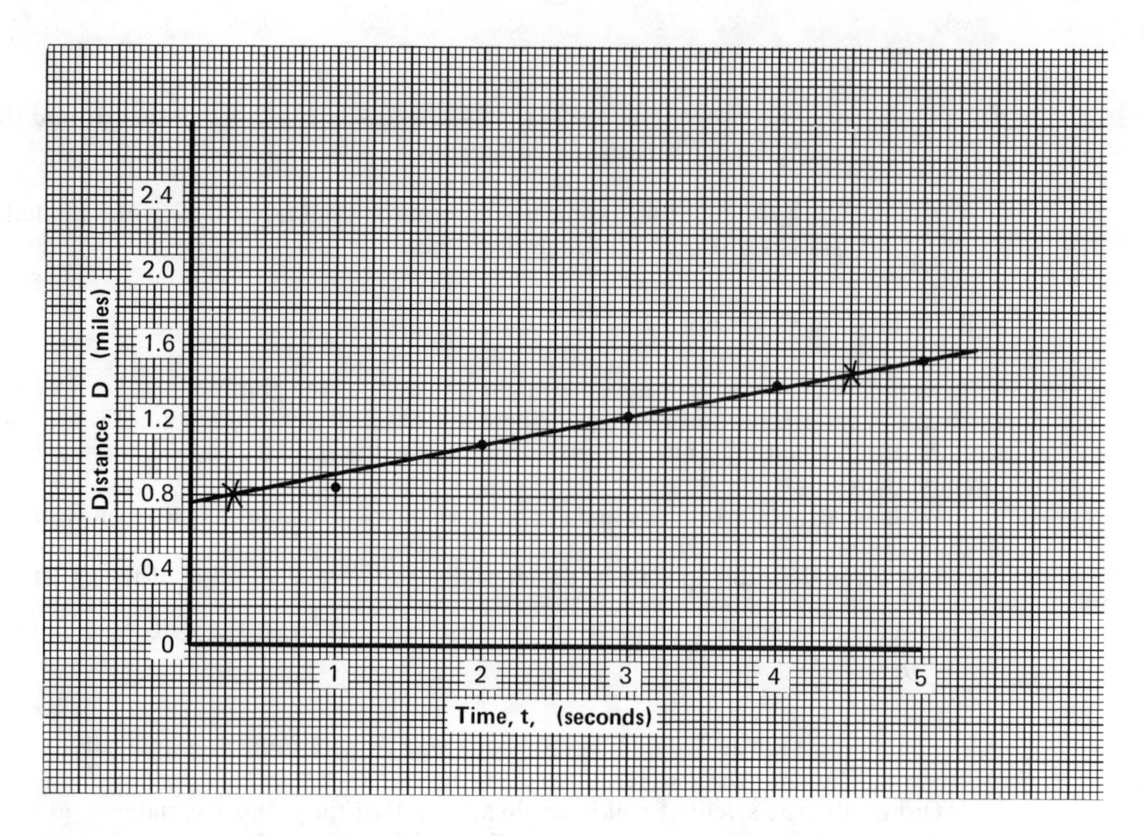

FIGURE 8

The data was recorded by observers at an auto race on a straight racetrack marked on a long sandy beach. The observers started recording time (t) and distance (D) when a car reached a steady speed on the track.

Notice that the data is almost but not quite ideal. What can we tell from this graph?

Evidently, this is a graph of a relationship like $y = mx + b$. Since D is graphed on the y-axis and t on the x-axis, the equation for this line is $D = mt + b$.

What is m, the slope?

$$m = \frac{D_2 - D_1}{t_2 - t_1}.$$

We have picked two data points and marked them on the graph with an X. Reading these, we get $D_2 = 1.44$ miles when $t_2 = 4.5$ seconds and $D_1 = 0.80$ miles when $t_1 = 0.30$ second. Then,

$$m = \frac{1.44 \text{ mi.} - 0.80 \text{ mi.}}{4.5 \text{ sec.} - 0.3 \text{ sec.}} = \frac{0.64 \text{ mi.}}{4.2 \text{ sec.}} = 0.15 \frac{\text{mi.}}{\text{sec.}}.$$

To determine b, read off the quantity for D when $t = 0$. It is about 0.75 miles. Thus,

$$D = \left(0.15 \frac{\text{miles}}{\text{second}}\right) t + 0.75 \text{ miles}$$

or $D = mt + b$ where $m = 0.15$ mi./sec. and $b = 0.75$ mi.

So far, the graph has provided an algebraic equation that enables calculation of the distance of the car from the start of the track at various times from the start of observing.

Next, it is noted that the unit for m is miles/second. This is a unit for speed and provides a clue to the meaning of m; in this case, it turns out to be the speed of the car. The y-intercept *b* has units of miles; the mile is a measure of distance. In this case, *b* turns out to stand for the distance from the start of the track to the point where observation started. Hence, the graph tells us that the car reached a steady speed of 0.15 miles/second at a distance of 0.75 miles from the start.

If you go on to study science, especially physics, you will come across other examples of such calculations and get more practice in interpreting the results. At this point, you should know how to calculate the slope of a straight line and how to write the equation for the line. While you should be able to interpret k in $y = kx$, you are not expected to interpret it in $y = mx + b$.

Here are some problems on graphical analysis.

6. Determine the slope of the line graphed for Problem 5, Section 7.3A.

7. Figure 9 shows a graph of masses and corresponding volumes of some pure silver bars measured at 20°C.

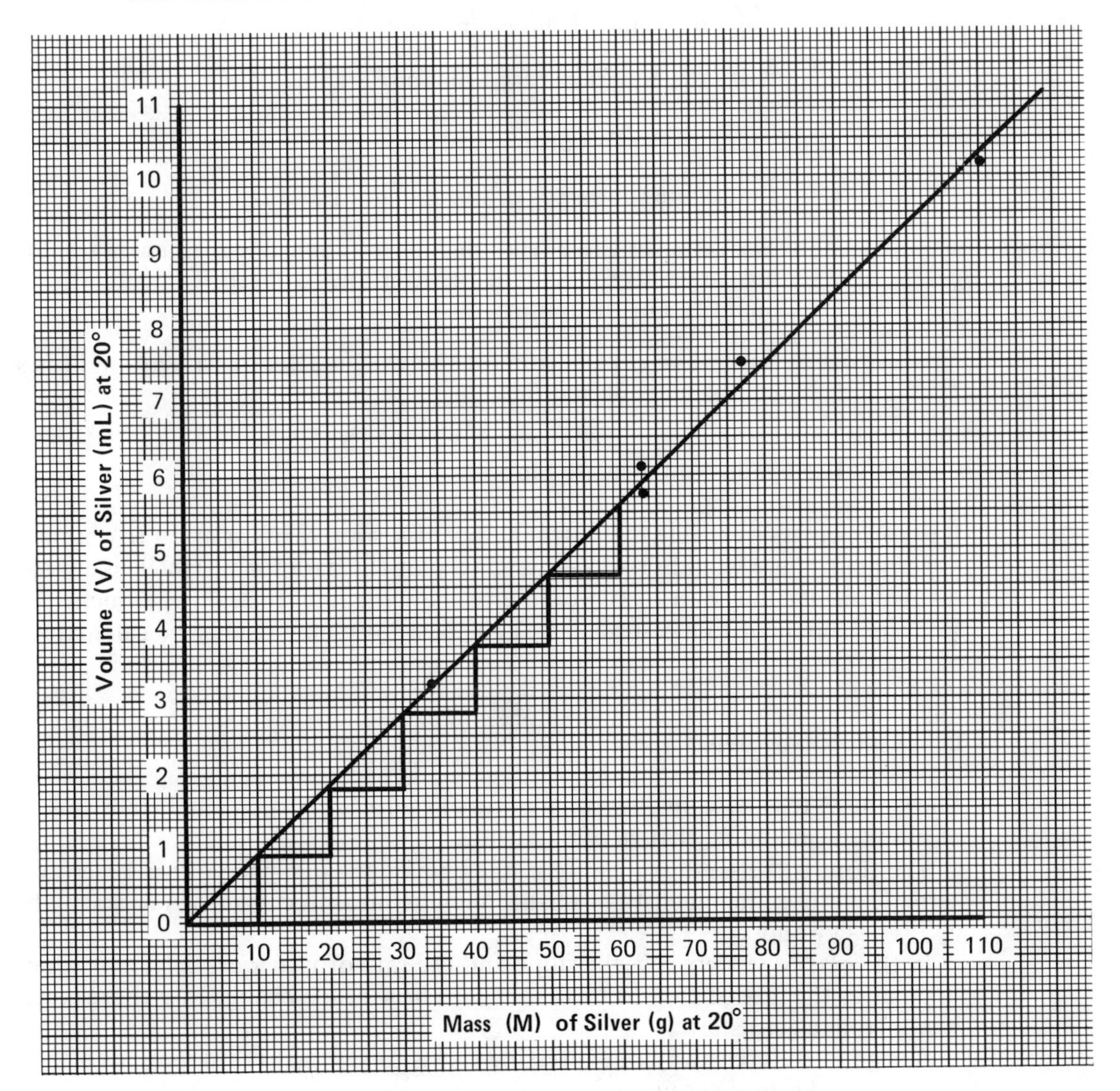

FIGURE 9

a. Calculate the slope using the general equation.

b. Write the equation for the graph; specify the values (quantities) for any constants.

c. State the rate in words.

d. What is the value of the intercept on the y-axis?

8. a. Calculate the slope of the line in the graph for Problem 20, Section 7.6C.

   b. From the graph, determine the y-intercept.

9. Use the graph you drew for Problem 22, Section 7.6C:

   a. Calculate the slope.

   b. Read the value of the y-intercept.

   c. Write the equation based on your graph.

10. Use Figure 10 for this problem:

   a. Determine the slopes of lines A, B, C, and D.

   b. The rates for these relationships have sensible real meanings. State the meanings in words for each of the four lines of best fit; include the respective quantities of the unitary rate.

   c. Suggest an explanation for the wide scatter of data around lines C and D as compared to A and B.

C. Why Use Graphs to Calculate Slopes?

If experimenters were able to control all possible sources of variation to obtain ideal data in an experiment, there wouldn't be any need to plot the data for variables likely to be related by a direct proportion. Every y/x would give the same rate. In real experiments, y/x may vary from measurement to measurement. The "best" y/x may readily be calculated from the line of best fit.

D. Concerning Slopes

The slope m tells how much *change* occurs in y for a *change* in x. It tells this as a quotient of numbers and units, that is, of quantities.

The slope is the same as the rate in the case of $y = mx$ since y equals zero when x equals zero. Therefore, the quantity of change may be measured from a zero start when $y_2 - y_1 = y_2$ and $x_2 - x_1 = x_2$.

In the case of $y = mx + b$, the slope and the rate of y to x are different when $b \neq 0$. The slope gives the rate of $(y - b)$ to x, so it tells how much change in y there is from a start of b, not from a zero start.

A slightly different way of expressing this is to say that the slope shows the number of units of change in y per unit of change in x.

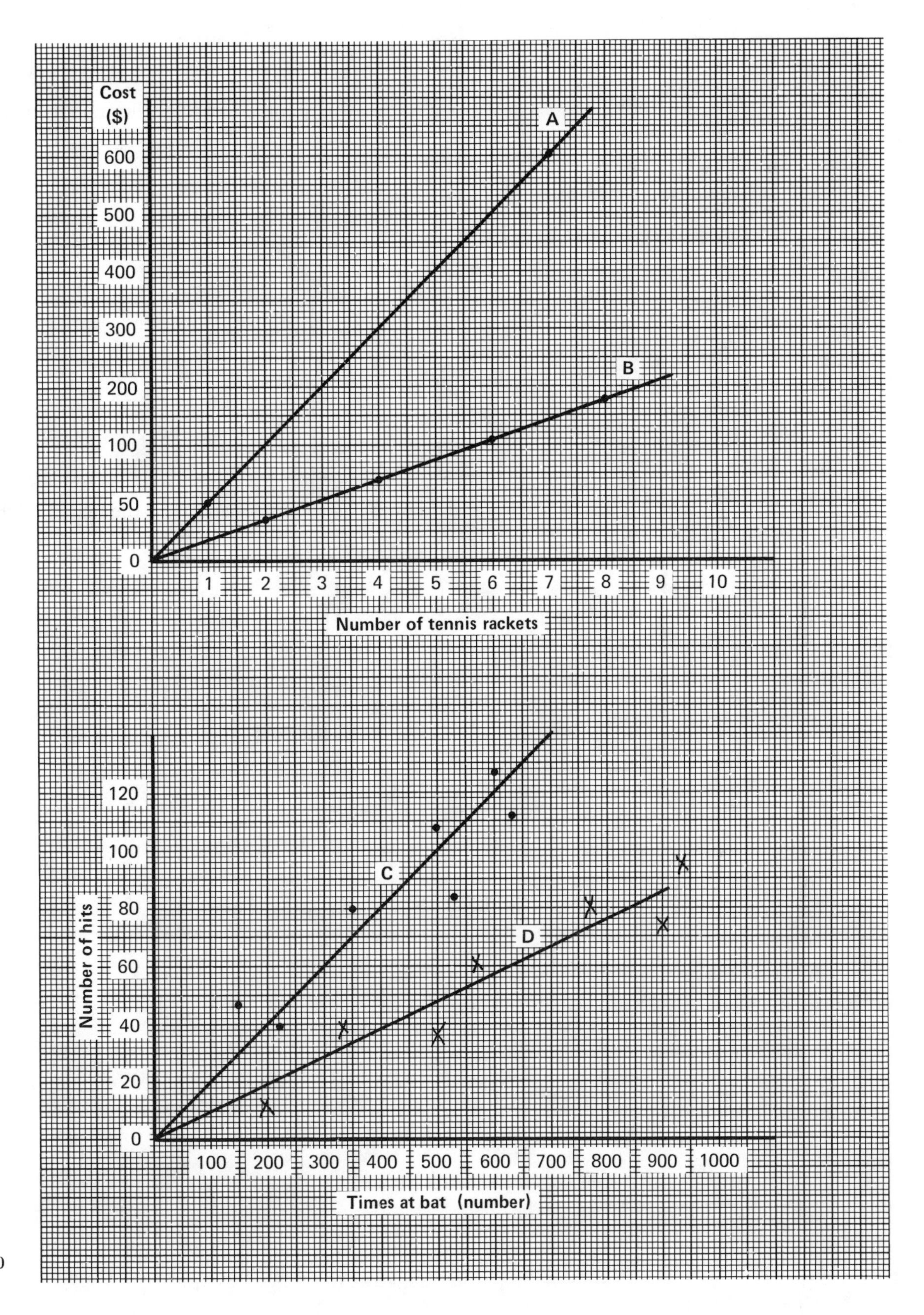
Cost
($)
600
500
400
300
200
100
50
0
A
B
1
2
3
4
5
6
7
8
9
10
Number of tennis rackets
120
100
80
60
40
20
0
Number of hits
C
D
100
200
300
400
500
600
700
800
900
1000
Times at bat (number)

FIGURE 10

This can be seen by examining what $y_2 - y_1$ and $x_2 - x_1$, in the expression

$m = \dfrac{y_2 - y_1}{x_2 - x_1}$, look like on a graph. We start with 2 points on a line whose paired coordinates are y_2, x_2 for one point and y_1, x_1 for the other point.

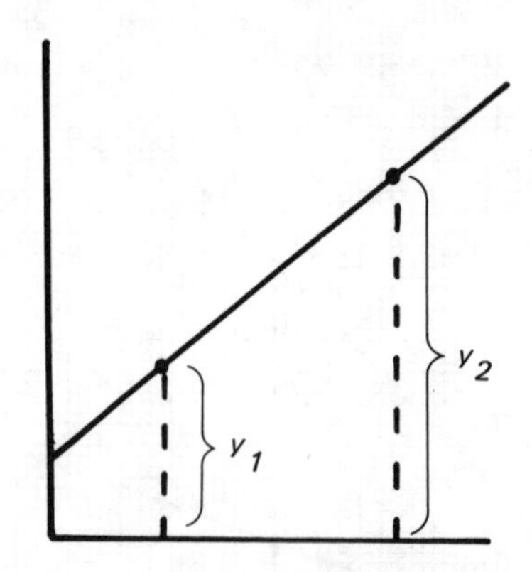

Now, y_1 is the height of the lower data point on the y-axis and y_2 is the height of the upper point on the y-axis.

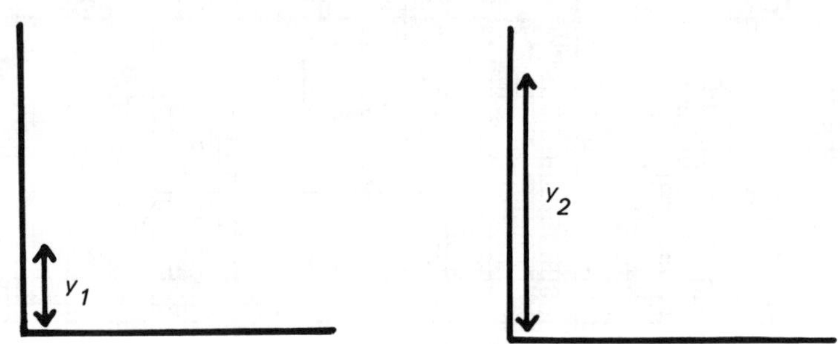

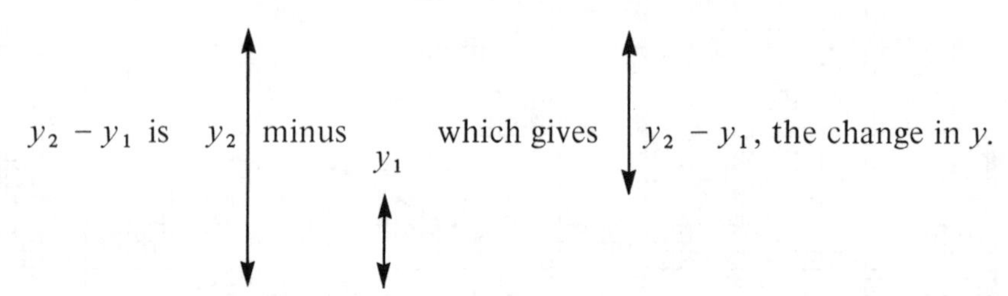

$y_2 - y_1$ is y_2 minus y_1 which gives $y_2 - y_1$, the change in y.

Likewise

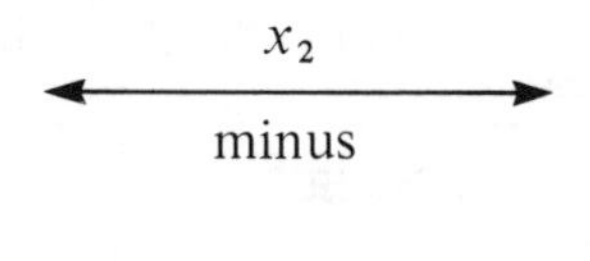

x_2

minus

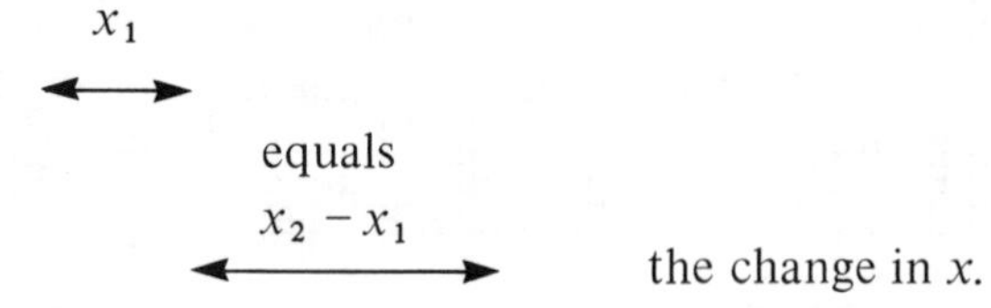

x_1

equals

$x_2 - x_1$ the change in x.

$y_2 - y_1 \quad \div x_2 - x_1$ equals m.

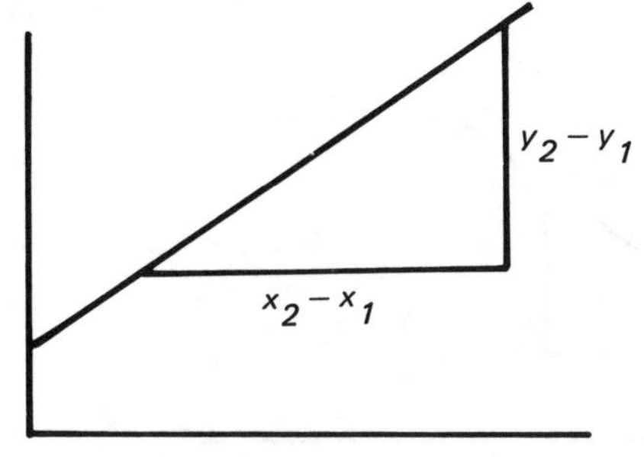

or

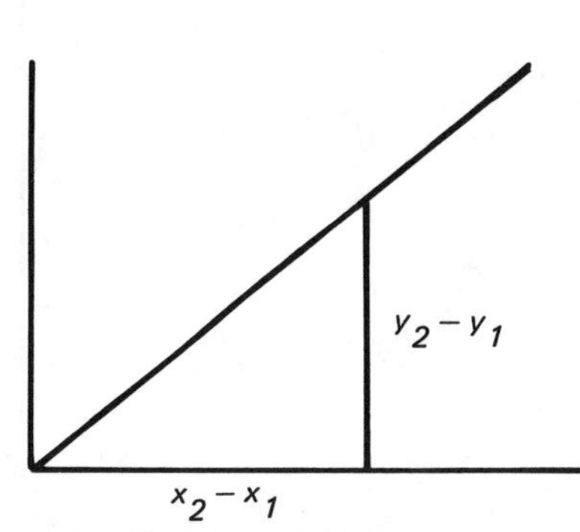

The slope gives the rate of *increase* of y to the corresponding *increase* in x.

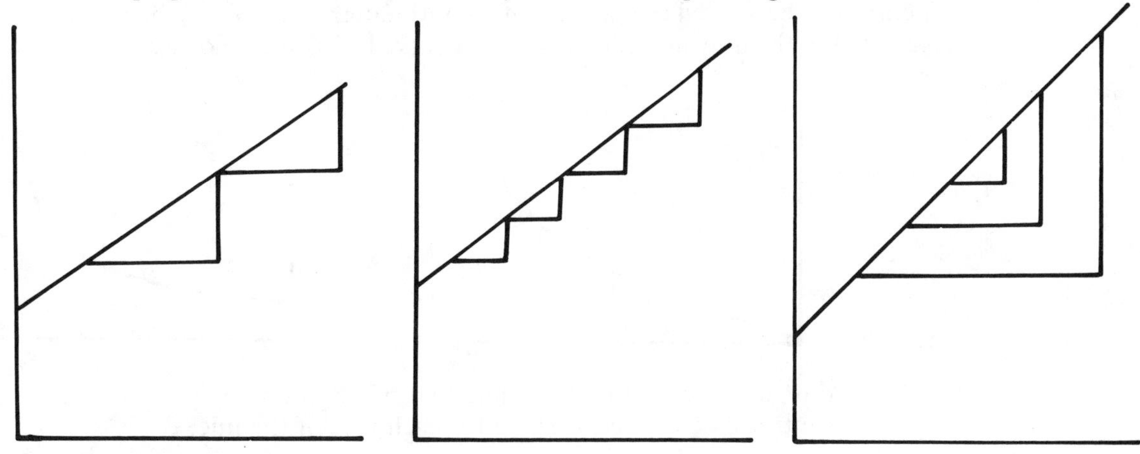

The slope is the same all along any straight line.

If two lines are drawn on the same graph, the rate of *change in* y with respect to *change in* x can be compared to show which changes more rapidly.

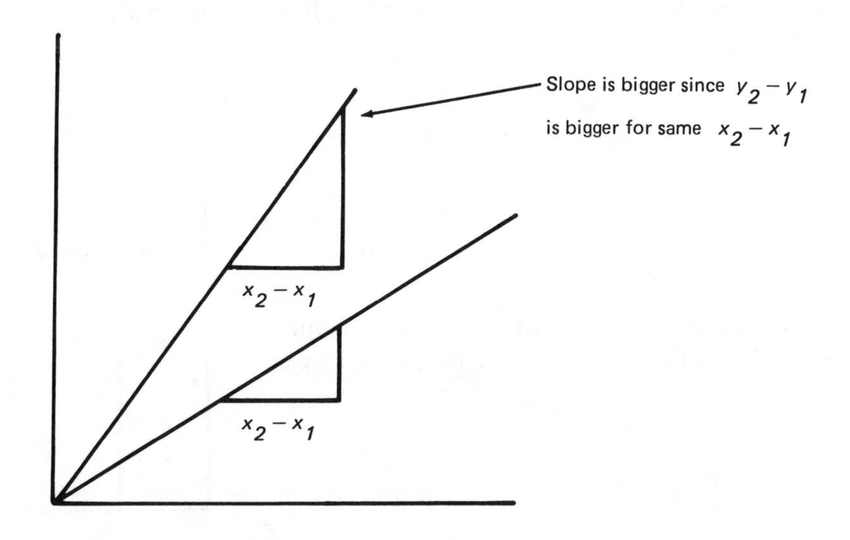

If a graph line bends, the two parts have different slopes.

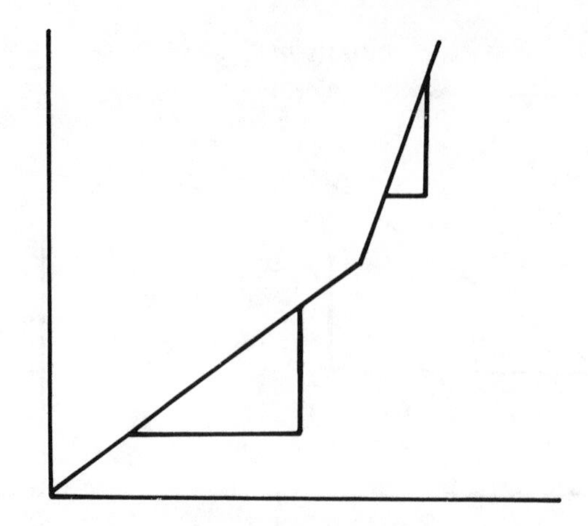

Only if the line is unbent, that is, straight, can all of it have the same slope.

A curve can be viewed as made up of a continuous series of tiny bends. They are exaggerated in the drawing below. Hence, a curve has a continuously changing slope.

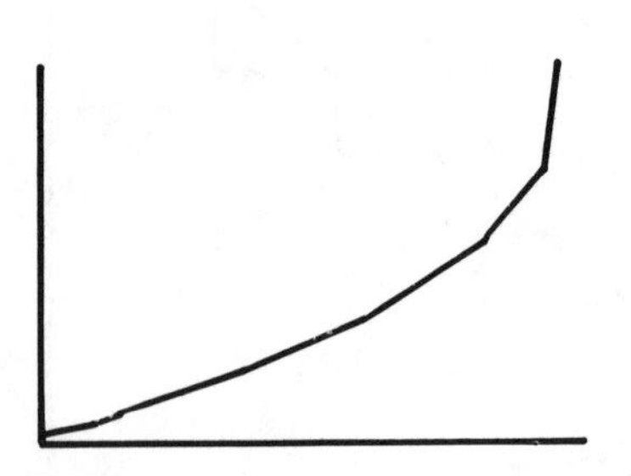

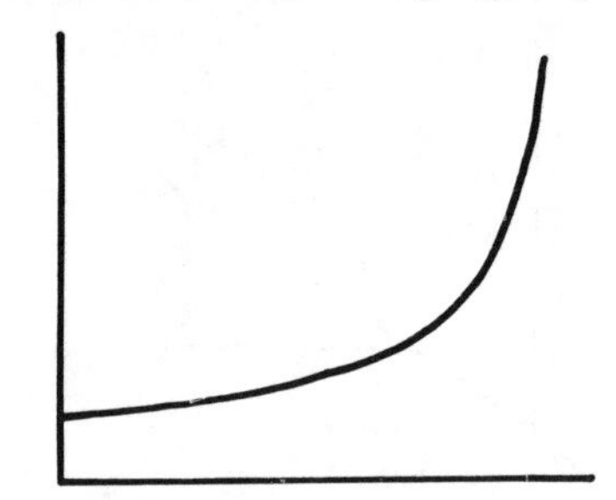

11. Construct a graph that has a line that bends at one point. Mark the scales with quantities. Calculate the slope for each part of the line.

12. On the graph you drew for Problem 11, mark off the lengths of $y_2 - y_1$ and $x_2 - x_1$ that you used to calculate the slope.

13. A graph has a horizontal line on it as shown below. What is the rate of change of y with respect to x?

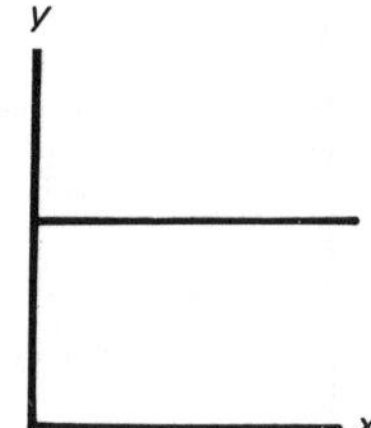

14. A graph has a vertical line on it as shown to the right. What is the rate of change of x with respect to y?

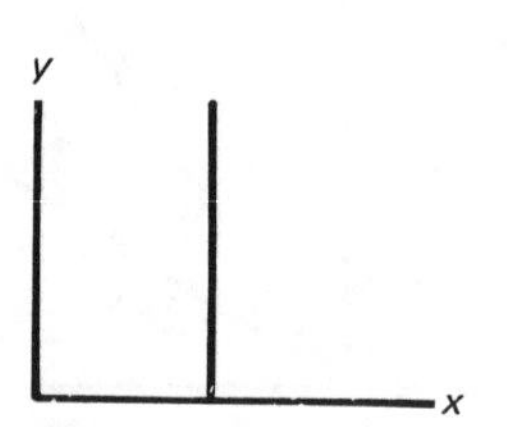

7.8 GRAPHS OF INVERSE PROPORTIONS

Unlike the direct proportion, the equation for an inverse proportion does not produce a straight line. The curve for $yx = k$ looks like this:

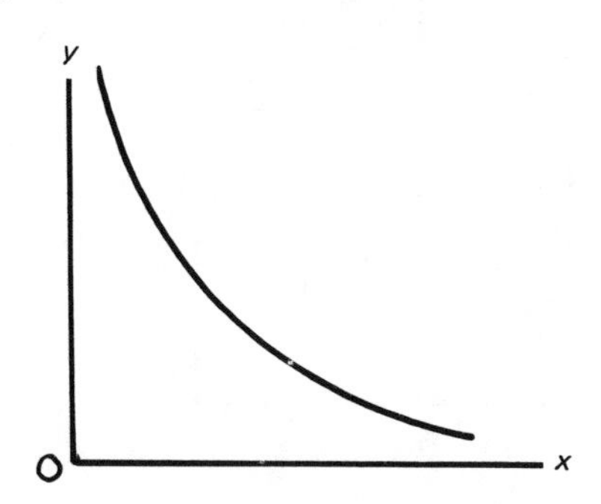

Thus, there is no slope calculation possible for the entire curve.

This curve can be recognized as showing an inverse relationship since y gets smaller as x increases. However, there are other inverse relationships that give curves of similar appearance. How can graphical analysis be used to show an inverse proportion given real data with considerable scatter around the ideal curve?

There is one way to graph inverse relations to show whether they are proportions; it is a neat trick. It depends on the fact that $yx = k$ can be rewritten as:

$$y = k \cdot \frac{1}{x} \quad \text{or} \quad \frac{y}{1/x} = k.$$

When rewritten this way, it can be seen that in an inverse proportion, *y and $1/x$ are **directly** proportional.*

Let's see how this works for three tables of inverse proportions.

1.

x	y	$1/x$
1	12	1
2	6	$\frac{1}{2}$
3	4	$\frac{1}{3}$
4	3	$\frac{1}{4}$

$yx = 12$

$\frac{y}{1/x} = 12$

2.

x	y	$1/x$
1	20	1
2	10	$\frac{1}{2}$
3	$6\frac{2}{3}$	$\frac{1}{3}$
4	5	$\frac{1}{4}$

$yx = 20$

$\frac{y}{1/x} = 20$

3.

x	y	$1/x$
1	5000	1
2	2500	$\frac{1}{2}$
3	1667	$\frac{1}{3}$
4	1250	$\frac{1}{4}$

$$yx = 5000$$
$$\frac{y}{1/x} = 5000$$

From the above, it can be seen that when $yx = k$, y is directly proportional to $1/x$. In that case, when y is graphed against $1/x$, the result will be a straight line going through the origin. Likewise, if x is graphed against $1/y$, the result will be a straight line going through the origin.

In the following, state what variables must be graphed to give a straight line going through the origin.

1. Length × width = area where area is constant.
2. Number bricks × length of a brick = length of the wall.
3. Number of people × time in hours = people-hours where people-hours is constant.

Can you extend this idea? Try to figure out what has to be graphed in each of the following to give a straight line through the origin. Remember that in each case you are seeking to arrange the variables into an expression of:

$$\frac{\text{Quantity}_2}{\text{Quantity}_1} = \text{constant.}$$

4. $cb = a$ where a is constant while c and b vary.
5. $y - a = kx$ where a and k are constants. (The answer is at the bottom of the page.)
6. $y = k\,(x - b)$ where b and k are constants.
7. $y = kx^2$.
8. $y = k\sqrt{x}$.
9. $y = (k/x) - a$.

Answer to 5: Plot $y - a$ against x.

10. $y = k/x^3$.

11. $y = (k/x^3) - 5$.

12. $y = k/\sqrt[3]{x}$.

13. $°F = \frac{9}{5}°C + 32°$.

14. Given data that shows that two variables are inversely but not proportionately related, suggest how to identify the relationship by use of a graph.

7.9 ALL KINDS OF RELATIONSHIPS

By now, you may have gathered that any kind of continuous relationship between two variables can be described in three ways:

1. By data.
2. By a graph.
3. By an equation.

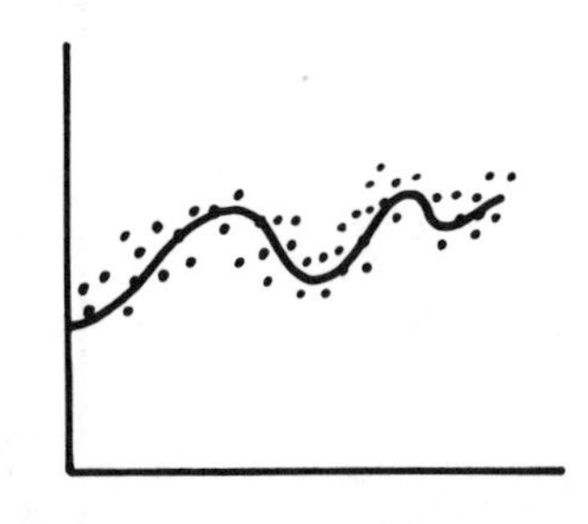

Each has its use. An equation can be found to match the curve drawn at the right from data but it may be a very complex equation, with several terms and varied exponents. If you are mathematically quite advanced, you may get a very precise mental picture from the equation of this complex curve and also be able to make calculations and derivations from the equation. For those of us who are not skilled mathematicians, the graph gives a mental picture as good as or better than the equation. For simpler curves, the equation may be more useful. The curve below, whose equation has not been studied here, still gives us considerable information. Among other things, it shows that as x increases, y increases, reaches a maximum and then decreases.

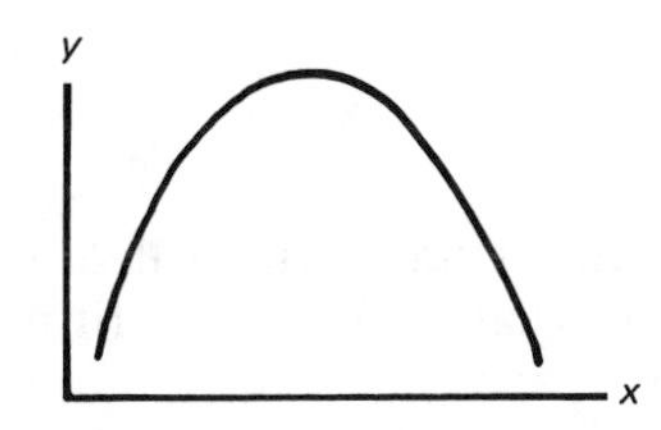

Describe as much as you can about the relationship between y and x in the following.

1.

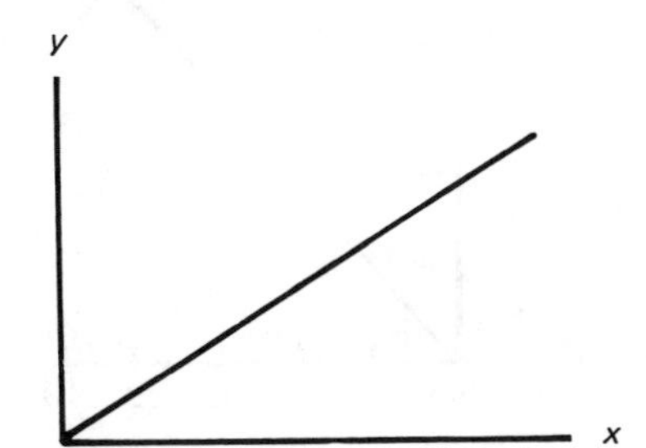

2.

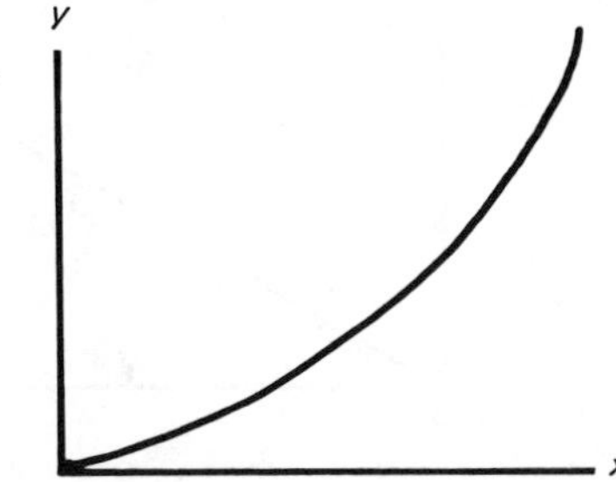

3.

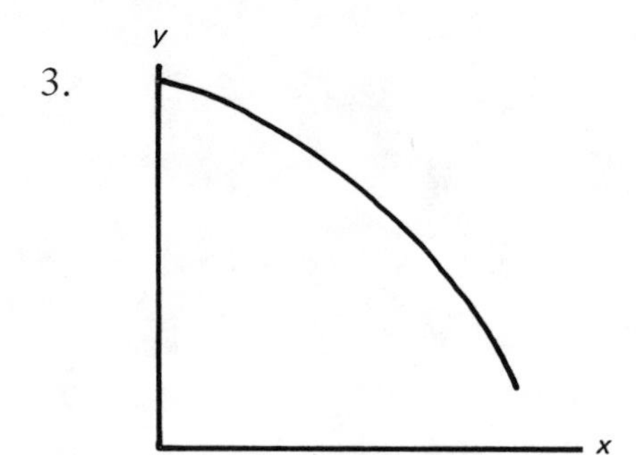

4.

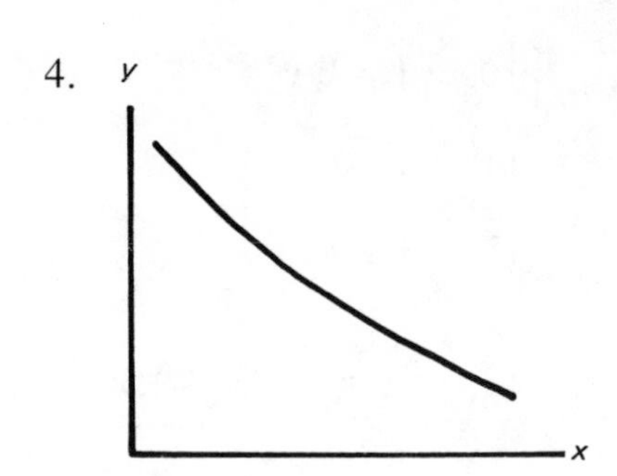

5.

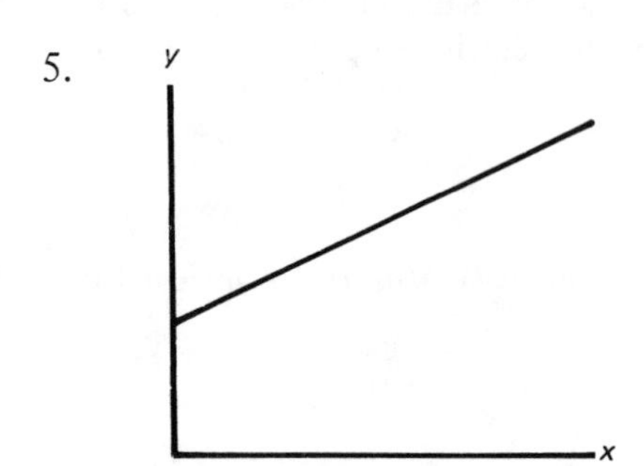

6.

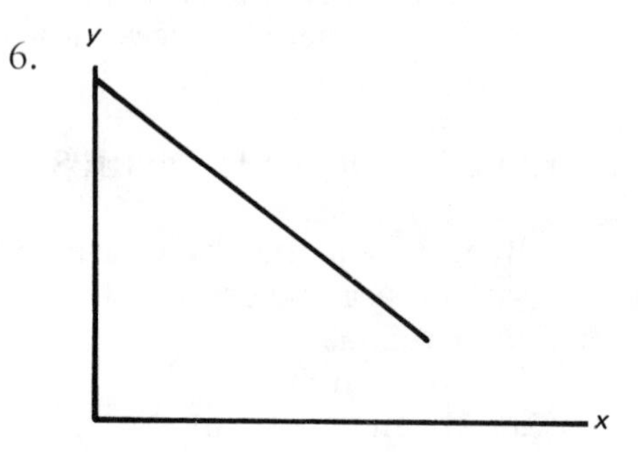

7.

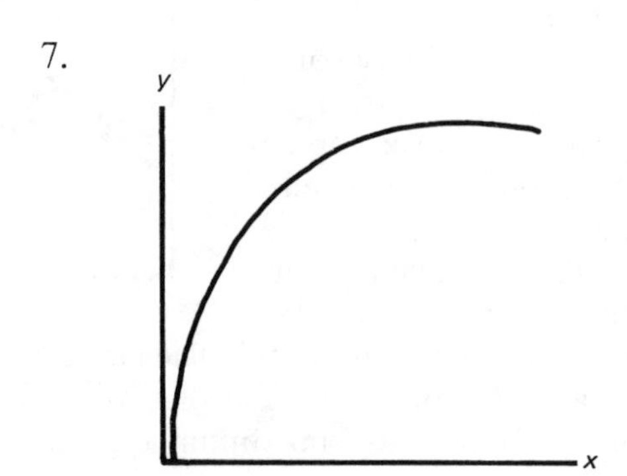

8.

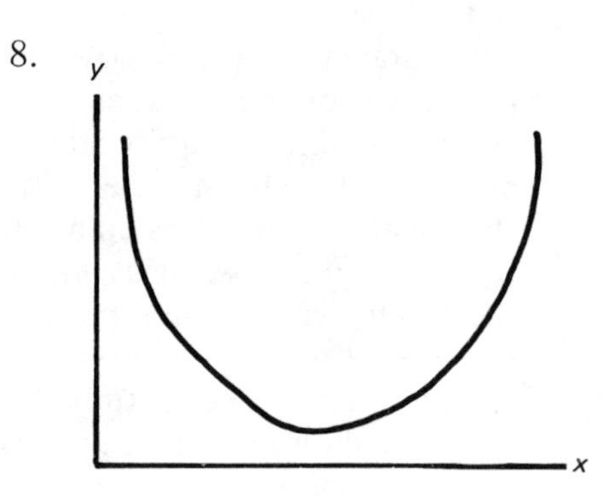

7.10 DON'T BE DECEIVED

Recall that in constructing a graph, the scale is selected by the designer of the graph. Given the same *y* scale, here is what a line of best fit might look like with three different scales.

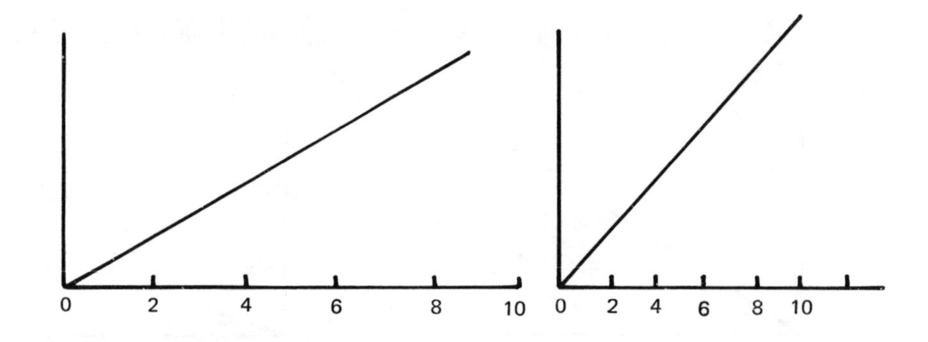

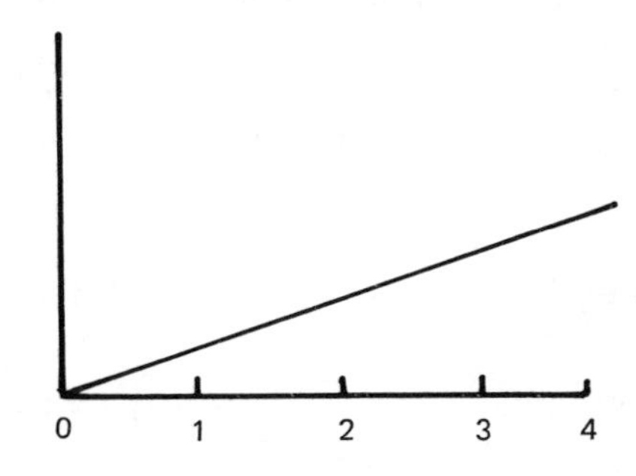

Has the data changed? No, it has not, yet the picture looks different for each graph. This points out a limitation for the use of a graph. Sometimes, one sees a graph in a magazine which is designed to show a very steep growth in a certain variable. Beware! You know that the scale can be manipulated to show the line rising very steeply.

Given the following data, graph it three ways, first so that it fills at least one half of an 8½″ × 11″ sheet, second with the x-axis scaled to half as long as it was before, and third with the x-axis scaled to one-fourth of the first graph.

x-axis	W (lbs.)	0	1	2	3	4	5
y-axis	L (in.)	0	2	4	6	8	10

This data shows the length (L) of a heavy spring as the weight (W) hung from it increases. Answer the following questions.

1. Do all three graphs show a direct proportion?
2. For each graph, calculate the slope.
3. What is the physical meaning of the slope?
4. Do the three graphs have different slopes? Why?
5. Write the equation for the first graph. Does it differ for each graph? Why?

DEFINITIONS

Dependent variable: The variable plotted along the vertical axis, or the set of first numbers of the ordered pairs, or the variable in an equation whose quantity is calculated after substitution for one or more of the other variables and constants in an equation.

Ideal data: Experimental data that agrees exactly with quantities calculated from an equation, or experimental data that falls precisely on the line of best fit.

Independent variable: The variable plotted along the horizontal axis, or the set of second numbers of the ordered pairs, or the variable whose quantity is substituted into an equation.

Line of best fit: A straight line or smooth continuous curve that passes as close as can be managed to all of the plotted points on a graph.

Quadrant: One of the four areas bounded on two sides by the x- and y-axes of a graph in the rectangular coordinate system.

Real data: Data obtained by measurement.

Slope of a line: Slope $= m = \frac{y_2 - y_1}{x_2 - x_1}$. It is the rate of change in y per change in x, $x_2 \neq x_1$.

ACTIVITY 7.1–Volume of One Penny

Purpose: To determine the volume of a penny.

Equipment: 28 new pennies, a 100 mL graduated cylinder.

Discussion: How much space does a penny occupy–what is its volume? One way to find out is to put the coin into a liquid such as water. The water will be displaced by the coin and rise up in the container. By measuring the difference in volume of the water before and after immersing the coin, the volume of the coin is determined.

By placing the coin into a 100 mL graduated cylinder, the volumes in millilitres can be read before and after immersing the coin.

Procedure: Fill a 100 mL graduated cylinder about half-way with water. If you have not already been instructed on how to read a meniscus, your teacher will tell you how to do this.

Read the volume of the water and record in the Report Sheet (1a).

Add 4 pennies to the water. Record (1b) and calculate the volume of the 4 pennies (1c).

Add 4 more pennies, repeat reading, record, and complete (2). Repeat until all 28 pennies are used, and record on the Report Sheet.

Calculations: First, graph your data. The variable on the x-axis is "Number of Pennies." The variable on the y-axis is "Volume of Pennies (mL)." Use ink for all of the graph except data points and line of best fit where pencil may be used. At the upper left, write your name. At the upper right, place a table of the quantities plotted. Plot each point with a small pencil dot; place a small circle around it to show that it is not a printer's blotch: ⊙ .

After all the points are plotted, draw a line of best fit. A length of black thread can be used to help you see where the line should go. There is one point on the graph to consider which may not actually be determined–the origin. If you have *no* pennies, is there any volume occupied? The answer is, of course, no. Hence, when the number of pennies is zero, the volume occupied is zero, and your line of best fit *must* pass through the origin.

After you have drawn the line of best fit, mark off a point near the beginning of your line and a point near the end (do *not* select data points) with a small cross (x). These are the points you will use to calculate the slope of your line. Call the paired quantities for the upper point y_2 and x_2 (record on Report Sheet). Call the paired quantities for the lower point y_1 and x_1; record. Calculate the slope m according to the equation: (record on Report Sheet).

$$m = \frac{y_2 - y_1}{x_2 - x_1}.$$

Hand in your graph with the Report Sheet. Answer the remaining questions.

ACTIVITY 7.2–Stress Test

Purpose: We know that pulse rate and exercise are directly related. Is this a proportional relationship as well? You will collect data to examine whether this is so. Care and patience are important in obtaining reliable data.

Equipment: A watch or clock with a second hand and a low chair or stool.

Procedure:

1. Sit quietly for a few minutes. Then take your pulse for 15 seconds and record under (1). Now step up onto the stool or chair and down again five times without stopping. Immediately count pulse beats for 15 seconds and record under (2).
2. Sit quietly until your pulse rate is back in the neighborhood of (1). Step onto the stool ten times. Immediately count pulse beats for 15 seconds and record under (3).
3. Repeat the above procedure for 15, 20, and 25 step-ups. Record on Report Sheet.
4. Answer remaining questions on the Report Sheet.

MODULE TWO

Answers to Selected Problems

CHAPTER ONE

Section 1.3

1. a, d
3. c, d
5. a. 2 or the reciprocal
   c. 2.6 or the reciprocal
7. a, d

Section 1.4

1. a. 15 tennis balls
   b. 336 cans
   c. $7740
3. 15 gallons
5. 1587 stitches
7. 15,120 g
9. 23 points

Section 1.5

A. 4. 2 inches or $\frac{1}{6}$ foot

Section 1.5 (continued)

B. 9. 25 cm^2
11. 6 miles/hour
13. 2 cm
15. 180 g · cm
17. 58.5¢/child

C. 19. feet
21. hours
23. inches
25. $\frac{\text{staples}}{\text{box}}$
27. $\frac{\text{newtons} \cdot \text{meters}^2}{\text{coulombs}^2}$

Section 1.6

1. a. 1 box/2000 staples
   c. 1 day/24 hours
2. a. 3 lbs./$1

Section 1.6 (continued)

3. a. $\frac{1/5}{1}$ or $0.2\,\frac{\text{hangars}}{\text{airplane}}$, $\frac{\text{5 airplanes}}{\text{hangar}}$

   c. $\frac{\text{44 g}}{\text{mole}}$, $\frac{\text{0.023 moles}}{\text{g}}$

   e. $\frac{3/4}{1}$ or $0.75\,\frac{\text{clips}}{\text{fastener}}$, $1.33\,\frac{\text{fasteners}}{\text{clip}}$

CHAPTER TWO

Section 2.2

1. length
3. time
5. volume
7. cost
9. density
11. volume
13. mass of bar per length of bar
15. number of pills per bottle
17. length in inches per same length in feet
19. heat energy (or Calories) per apple

Section 2.3

1. $y = 1\frac{1}{4}$

3. $y = 5$; the constant is $\frac{1}{8}$

Section 2.5

1. a. C and I
   c. (1) $90
   e. $45
   g. P is constant for each problem, but changes from problem to problem.
3. a. $d = 19.3$ g/mL
   c. $v = 0.074$ mL

Section 2.6

1. b and d violate the rule.

3.

	y	a	x
a.	litres	litres/qt.	quarts
b.	cm	cm/sec	sec
c.	cans	cans/case	cases

Section 2.7

1. y changes by a factor of 8.
3. y changes by a factor of 0.02.
5. y changes by same factor.
7. s becomes $\frac{1}{10}$ as big.
9. No

Section 2.8

A. 1. Yes
3. No
5. No

B. 7. Yes
9. Yes
11. No
13. No
15. Yes
17. Yes

C. 19. 30
21. Doubles
23. Doubles
25. y^2 changes by the same factor.
27. zx

CHAPTER THREE

Section 3.1

3. $C = 100A$

Section 3.3

1. C = number of closet doors
   N = number of rooms
   k = number of closet doors per room
   $C = kN$
3. I = number of innings
   N = number of games
   k = number of innings per game
   $I = kN$
5. V = volume of paint
   a = area to be painted
   k = volume of paint per area
   $V = ka$

Section 3.3 (continued)

7. m = mass of oxygen
   v = volume of oxygen
   k = mass per volume of oxygen
   $m = kv$

Section 3.4

1. k = distance per time
   t = time
   D = distance
   $D = kt$
3. k = mass of solute (or solid) per volume of solution
   m = mass of solute
   v = volume of solution
   $m = kv$
5. k = length in inches per same length in feet
   I = length in inches
   F = same length in feet
   $I = kF$
7. k = heat energy per item
   H = heat energy
   I = number of items
   $H = kI$

Section 3.5

A. 2. 0.00333 hr./Cal.

B. 3. Not a proportion.
   5. a. $P = aA$ $\quad$ P = volume paint
         $A = kP$ $\quad$ A = area
         a = constant = volume/area
         k = constant = area/volume
      b. 1.96 gallons
   7. a. Y = yd.^{2} of area
         F = ft.^{2} of area

$$k = \text{constant} = \frac{\text{yards}^2}{\text{feet}^2} \qquad Y = kF$$

$$a = \text{constant} = \frac{\text{feet}^2}{\text{yards}^2} \qquad F = aY$$

      b. 41.0 yards^2
      c. 270 feet^2
   9. F = mass fudge
      C = energy

$$a = \frac{\text{mass fudge}}{\text{energy}} \qquad F = aC$$

$$k = \frac{\text{energy}}{\text{mass fudge}} \qquad C = kF$$

Section 3.5 (continued)

11. a. W francs, t days
    b. Yes
    c. Units must be equal on both sides of the equation, and be those correctly used for such a measurement.

Section 3.6

1.

Distance (miles)	Time (sec.)
2	9
4	18
6	27

d = distance
t = time
k = constant = distance/time = $0.22 \dfrac{\text{miles}}{\text{sec.}}$

$d = kt$

3. m = mass air
   v = volume

   $$d = \text{constant} = \frac{\text{mass}}{\text{volume}} = 1.2\,\text{g/L}$$

   $m = dv$
5. H = mass hydrogen
   O = mass oxygen

   $$k = \text{constant} = \frac{\text{mass hydrogen}}{\text{mass oxygen}} = 0.125$$

   $H = kO$

CHAPTER FOUR

Section 4.2

F. 1. 4.2
   3. 1409 pills
   5. 2.6 L

Section 4.3

1. Inverse proportion
3. Inverse proportion
5. Neither

Section 4.4

B. 1. a. $\dfrac{1}{25}$ b. 25 c. $\dfrac{10}{3}$ d. $\dfrac{1}{5}$
   3. a, b, d, f

Section 4.5

1. a. $\frac{96}{24} = 4; \frac{1/4}{1} = \frac{1}{4}$

   b. $\frac{96}{8} = 12; \frac{1/4}{3} = \frac{1}{12}$

3. a. $\frac{6}{48} = \frac{1}{8}; \frac{4}{1/2} = 8$

   b. $\frac{6}{24} = \frac{1}{4}; \frac{4}{1} = 4$

Section 4.6

1. b
3. d
5. b
7. d

Section 4.7

1. a. L = length
   W = width
   k = constant = length × width
   $LW = k$

   b. 30 feet

   c. Area in square feet

3. a. d = diameter
   l = length
   k = constant = diameter × length
   $dl = k$

   b. 8.7 cm

5. a. P = quantity (or number) of pearls
   d = diameter of pearls
   k = constant = diameter × quantity of pearls
   $Pd = k$

   b. 12.5 pearls, rounded to 12 pearls

Section 4.8

1. p = perimeter
   l = length of long side
   w = width of short side
   $p = 2l + 2w$
3. P = quantity of pieces
   r = pieces right-side up
   w = pieces wrong-side up
   $P = r + w$

Section 4.8 (continued)

5. L = length of chain = constant
   l = length per link
   q = quantity of links
   $L = lq$
7. c = length of chain
   l = total length of links
   k = constant length of clasp
   $c = l + k$

CHAPTER FIVE

Section 5.2

1. When density is constant, mass and volume are directly proportional to each other. As mass increases by any factor, the volume occupied by the object increases by the same factor. When the volume is held constant, mass and density are directly proportional to each other. As the density of an object increases by any factor, the mass that fills the same volume increases by the same factor.

Section 5.4

A. 1. a. Direct proportion; v and t; v and a
Inverse proportion; a and t
c. Direct proportion; squops and swindlers; squops and jiggers
Inverse proportion; jiggers and swindlers

2. a. For $3t$, $\frac{1}{3}a$; for $3v$, $3a$

c. For $3v$, $\frac{1}{3}d$; for $3m$, $3d$

B. 3. Inverse proportion
5. Direct proportion
7. a. Direct proportion
b. Direct proportion
c. Inverse proportion
9. a. Inverse proportion
b. Direct proportion

Section 5.5

A. 1. x^2
3. $\sqrt{ac}$
5. $\frac{b}{x}$

Section 5.5 (continued)

B. 7. ay

9. a. 4 b. 4 c. 16 d. 64 e. a^2

11. Becomes 9 times as great

13. a. $\frac{1}{4}$ b. $\frac{1}{9}$ c. $\frac{1}{100}$ d. $\frac{1}{a^2}$

15. a. $\frac{1}{4}$ c. 4 e. 100

17. a. Direct proportion
 c. Direct proportion
 e. Inverse proportion

Section 5.6

A. 1. A = Alice's earnings
 B = Belle's earnings
 b = \$50 = constant
 $A = B + b$

3. S = 16 slices = total slices
 $S = E + L$

5. P = 100 yards = perimeter
 L = length of side
 W = width of side
 $P = 2L + 2W$

8. 20°C = 68°F (from scale). To calculate for 40°C: $°F = (\frac{9}{5} \times 40°) + 32° = 104°$

 Next, calculate °F – 32° and compare: 68°F – 32°F = 36°F
 104°F – 32°F = 72°F
 $\frac{72°F}{36°F} = 2$

Section 5.7

1. a. d_1 = 55 miles, d_2 = 110 miles
 t_1 = 1 hr., t_2 = 2 hrs.

 $$\frac{d_1}{t_1} = \frac{d_2}{t_2}$$

 c. p_1 = 260 balls pitched
 p_2 = 100 balls pitched
 h_1 = 65 balls hit
 h_2 = 25 balls hit

 $$\frac{p_1}{h_1} = \frac{p_2}{h_2}$$

3. Inverse

Section 5.7 (continued)

5. (4) $t_1 = 1$ hr., $d_1 = 6$ miles; $t_2 = 3$ hrs., $d_2 = 18$ miles

$\frac{t_2}{t_1} = \frac{d_2}{d_1}$ or $\frac{t_1}{d_1} = \frac{t_2}{d_2}$

(6) $n_1 = 20$ beads, $l_1 = 5$ cm; $n_2 = 200$ beads; $l_2 = 50$ cm

$\frac{n_1}{l_1} = \frac{n_2}{l_2}$ or any rearrangement of this (or use $n_1 = 10$ beads, $l_1 = 2.5$ cm)

CHAPTER SIX

1. $\frac{\$3.45}{\text{rose}}$

3. a. B b. A c. C d. B e. C f. C

5. a. W = weight in pounds
M = mass in kg
$k = 2.2$ lbs./kg = mass in pounds per mass in kilograms
$W = kM$

c. L = length
C = cost
$k = \frac{1 \text{ yard}}{\$13}$ = length per cost
$L = kC$

7. W = weight luggage
P = quantity of passengers
$k = \frac{100 \text{ lbs. baggage}}{6 \text{ passengers}}$ = quantity of baggage per passenger
$W = kP$

9. a. Quadruples

c. Becomes 5 times as big

e. Becomes $\frac{1}{10}$ as big

g. 100

11. a. Increases. F and a are directly proportional.

c. Acceleration of truck is 1/3 that of a small car since mass and acceleration are inversely proportional.

13. a. d = distance
s = speed
k = constant time
$d = ks$

c. N = blocks/box
L = length of box
k = length of row
$NL = k$

CHAPTER SIX (continued)

15. 4 days

17. $1\frac{2}{3}$ weeks

19. 0.21 litre

21. a. $st = k$
    b. $s = kt^2$

23. Becomes 9 times as great

25. a. Quadruples
    b. 16 times as big

CHAPTER SEVEN

Section 7.2

1. 1, 16
3. 1.5, 5
5.

x	y
1	16
1.5	15
2.5	12
5.8	14.8
7.0	8

Section 7.3

A. 1. 98 g
   3. a. 97.8 g b. 117.6 L
      c. Same result but less precisely. Larger graph is needed to estimate to same number of digits.
   5. a.

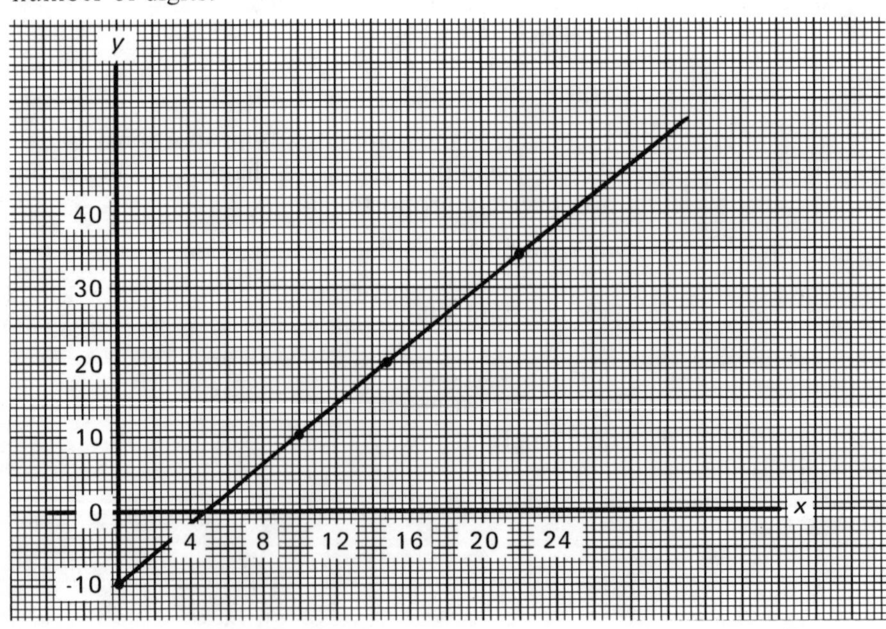

Section 7.3 (continued)

b. −6
c. 13

B. 7. b
9. d
11. c
13. a.

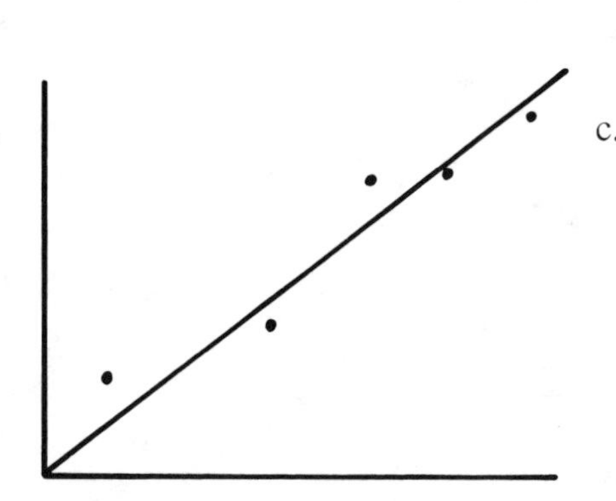

c.

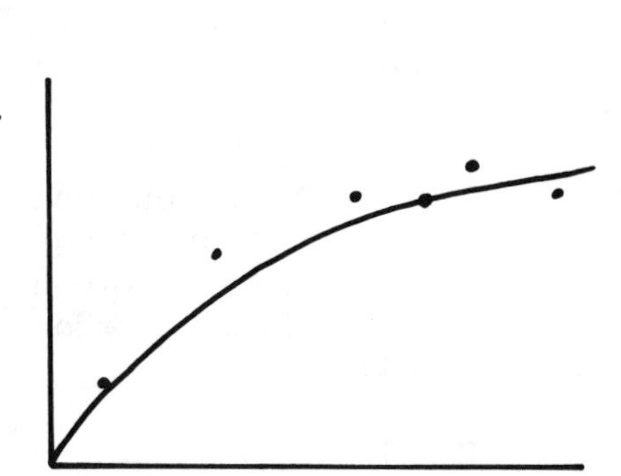

e.

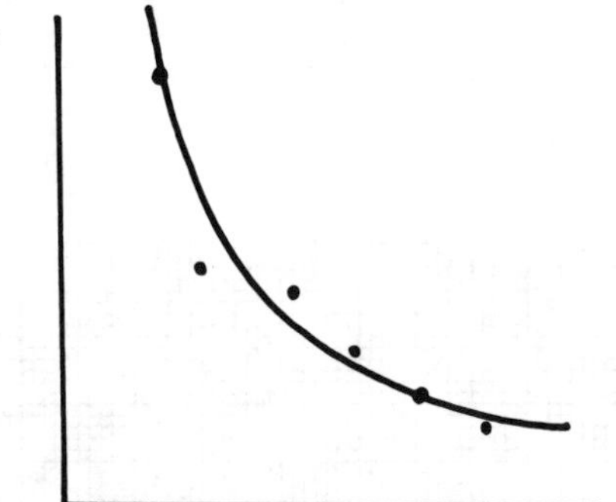

g.

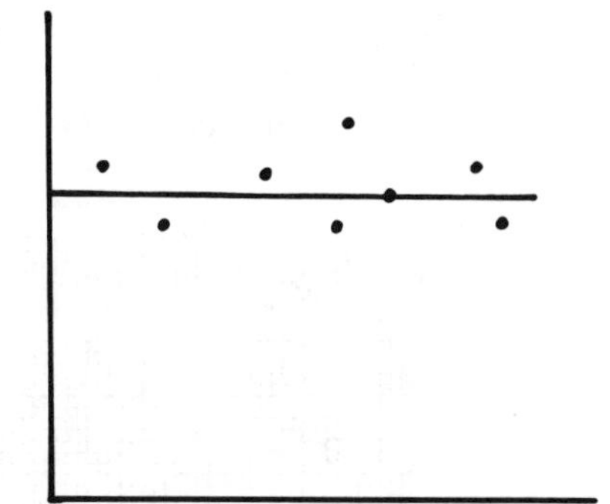

i. No relationship

Section 7.4

3. (7) Direct relationship
(9) Inverse relationship
(11) Inverse relationship
(13) a. Direct
c. Direct
e. Inverse
g. Neither
i. Neither

Section 7.5

A. 1. x, number of seeds
y, number of seedlings
3. x, passage of time
y, change in height

Section 7.5 (continued)

B. 6, 7, 9, 10

C. For the x-axis, (c) because the data is spread over the length of the axis and because it is scaled consistently.

Section 7.6

A.
1. $y = kx,\ k = 5$
3. Not a proportion
5. Not a proportion
7. Not a proportion
9. Not a proportion
11. c, d, f

B.
13. 0, yes
15. Yes, straight line going through the origin
17. Yes

C.
19. $y + b$
21. y intercept = 6

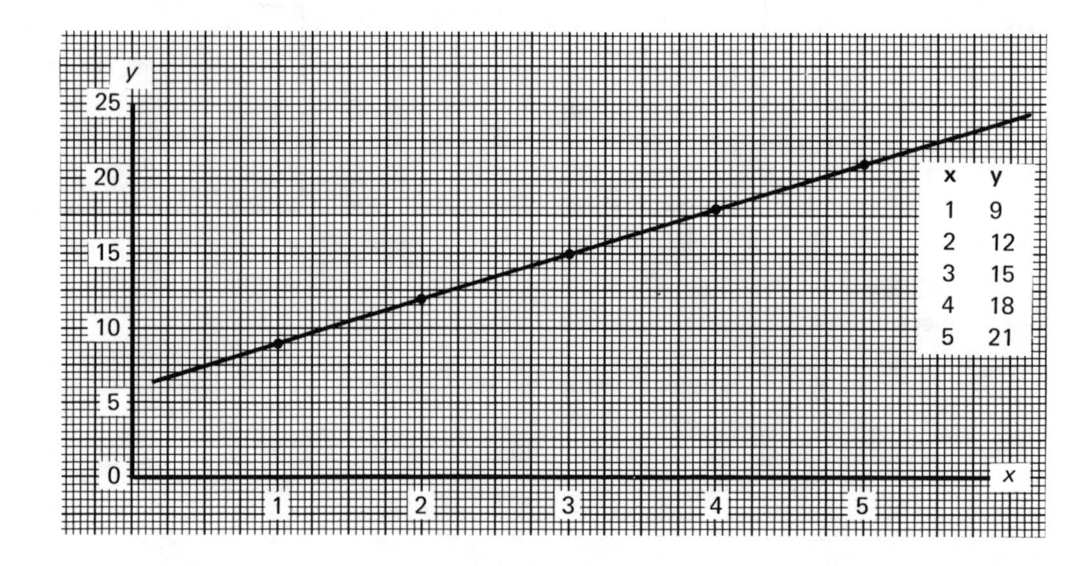

23. For $y = 5x - 2$, $y + 2$
For $y = 3x + 6$, $y - 6$
For $°F = \frac{9}{5}°C + 32°$, $°F - 32°$

25. It is the constant, b, in $y = kx + b$. It is the quantity for y when x equals zero.

Section 7.7

A. 1. Invariant rate should be between 0.38 and 0.40 g salt per mL water. It differs only because of estimations in readings; hence, it is invariant.

3. a. 5   b. 16   The graph reading should be the same allowing for any differences due to estimation.

B. 7. a. $\dfrac{0.096 \text{ mL silver}}{\text{g silver}}$

b. $V = kM$ where k is the volume per mass

c. Each gram of silver occupies 0.096 mL

d. Zero

9. a. 1.8°F/°C

b. 32°F

c. °F = k°C + 32°F where

$$k = \frac{\text{number of degrees Fahrenheit}}{\text{Celsius degree}} = \frac{1.8°\text{F}}{°\text{C}}$$

D. 13. Zero

Section 7.8

1. Length and reciprocal of width or width and reciprocal of length
3. Number of people and reciprocal of time or the converse
7. y and x^2
9. $(y + a)$ and $\dfrac{1}{x}$
11. $(y + 5)$ and $\dfrac{1}{x^3}$
13. (°F − 32°) and °C

Section 7.9

1. Direct proportion
3. Inverse relation
5. Direct relation
7. Direct relation; as x increases, the increase in y becomes smaller and then even smaller

Section 7.10

1. Yes
3. Rate of stretch of spring in response to weight
5. $W = kL$. Same for all.